Applied Categorical Data Analysis

WILEY SERIES IN PROBABILITY AND STATISTICS
TEXTS AND REFERENCES SECTION

Established by WALTER A. SHEWHART and SAMUEL S. WILKS

A complete list of the titles in this series appears at the end of this volume.

Applied Categorical Data Analysis

CHAP T. LE, PH.D.
Professor of Biostatistics, School of Public Health
Director of Biostatistics, Cancer Center
University of Minnesota

A Wiley-Interscience Publication
JOHN WILEY & SONS, INC.
New York · Chichester · Weinheim · Brisbane · Singapore · Toronto

This text is printed on acid-free paper. ♾

Published simultaneously in Canada.

Library of Congress Cataloging-in-Publication Data:

Le, Chap T., 1948–
Applied categorical data analysis / Chap T. Le.
p. cm. – (Wiley series in probability and statistics. Texts and references section)
"A Wiley-Interscience publication."
Includes bibliographical references and index.
ISBN 0-471-24060-5 (pbk.: alk. paper)
1. Multivariate analysis. I. Title. II. Series.
QA278.L39 1998
519.5'35–dc21

98-14782
CIP

To MinhHa, Mina, and Jenna with love.

Contents

Preface

This book is intended to meet the needs of practitioners and students in applied fields for a single, fairly thin volume covering major, updated methods in the analysis of categorical data. It is written for the training of graduate students in epidemiology, environmental health, and other fields in public health, as well as for biomedical research workers. It is designed to offer sufficient details to provide a better understanding of the various procedures as well as the relationships among different methods. In addition, the level of mathematics has been kept to a minimum. As a book for students in applied fields and as a referenced book for practicing biomedical research workers, *Applied Categorical Data Analysis* is application oriented. It introduces applied research areas, with a large number of real-life examples and questions, most of which are completely solved; samples of computer programs are included.

I would like to express my thanks to colleagues for their extensive comments and suggestions; the comments from many of my graduate students at the University of Minnesota also have been most helpful. At the University of Minnesota, we offer a three-quarter course sequence for students in Epidemiology and Environmental and Occupational Health, and a few other graduate programs in the Academic Health Center. The last quarter of this sequence, devoted to the study of techniques used in analyzing categorical data, and the contents herein, represent my multiyear effort in teaching that course.

CHAP T. LE

Edina, Minnesota

CHAPTER 1

Introduction

The purpose of most research is to assess relationships among a set of variables, and choosing an appropriate statistical technique depends on the type of variables under investigation. Suppose we have a set of numerical values for a variable:

(i) If each element of this set may lie only at a few isolated points, we have a *discrete* or *categorical* data set. In other words, a categorical variable is one for which the measurement scale consists of a set of *categories*; examples are race, sex, counts of events, or some sort of artificial grading.

(ii) If each element of this set may theoretically lie anywhere on the numerical scale, we have a *continuous* data set. Examples are blood pressure, cholesterol level, or time to a certain event such as death.

This text focuses on the analysis of categorical data and the eventual aim concerns *multivariate* problems when at least three variables are involved. The first section of this chapter shows a simple example of *real* problems to which some of the methods described in this book can be applied. This example shows a potential complexity when data involve more than two variables with a phenomenon known as effect modification. We will return to this example later when illustrating some methods of analysis for categorical data in

Chapters 2, 3, and 4. The second section briefly reviews some likelihood-based statistical methods to be used in subsequent chapters with various regression models. The final section summarizes special features of this text, its objectives, and for whom it is intended.

1.1. A PROTOTYPE EXAMPLE

Many research outcomes can be classified as belonging to one of two possible categories: Presence and Absence, Non-white and White, Male and Female, Improved and Not-improved. Of course, one of these two categories is usually identified as of primary interest, for example, Presence in the Presence and Absence classification, Non-white in the White and Non-white classification. We can, in general, relabel the two outcome categories as Positive (+) and Negative (−). An outcome is positive if the primary category is observed and is negative if the other category is observed, and health decisions are frequently based on the proportion of positive outcomes defined by

$$p = \frac{x}{n},$$

where x is the number of positive outcomes from observations made on n individuals; $0 \leq p \leq 1$. Proportion is a number used to describe a group of individuals according to a dichotomous characteristic under investigation and the example below provides an illustration of its use in the health sciences.

Comparative studies are intended to show possible differences between two or more groups. Data for comparative studies may come from different sources, with the two fundamental designs being retrospective and prospective. Retrospective studies gather past data from selected cases and controls to determine differences, if any, in the exposure to a suspected risk factor. They are commonly referred to as *case-control studies*. Cases of a specific disease are ascertained as they arise from population-based registers or lists of hospital admissions, and controls are sampled either as disease-free individuals from the population at risk, or as hospitalized patients having a diagnosis other than the one under study. The advantages of a retrospective study are that it is economical and it is possible to obtain answers to research questions relatively quickly because the cases are already available. Major limitations are due to the inaccuracy of the exposure histories and uncertainty about the appropriateness of the control sample; these problems sometimes hinder retrospective studies and make them less preferred than prospective studies. The following example introduces a retrospective study concerning occupational health.

Example 1.1. A case-control study was undertaken to identify reasons for the exceptionally high rate of lung cancer among male residents of coastal Georgia (Blot et al., 1978). Cases were identified from these sources:

(a) diagnoses since 1970 at the single large hospital in Brunswick,
(b) diagnoses during 1975–76 at three major hospitals in Savannah, and
(c) death certificates for the period 1970–74 in the area.

Controls were selected from admissions to the four hospitals and from death certificates in the same period for diagnoses other than lung cancer, bladder cancer, or chronic lung cancer. Data are tabulated separately for smokers and nonsmokers as follows:

Smoking	Shipbuilding	Cases	Controls
No	Yes	11	35
	No	50	203
Yes	Yes	84	45
	No	313	270

The exposure under investigation, "Shipbuilding," refers to employment in shipyards during World War II.

In an examination of the smokers in the above data set, the numbers of people employed in shipyards, 84 and 45, tell us little because the sizes of the two groups, cases and controls, are different. Adjusting these absolute numbers for the size of the respective group, we have

(i) for the controls,

$$\text{Proportion of exposure} = \frac{45}{315}$$
$$= .143 \text{ or } 14.3 \text{ percent;}$$

(ii) for the cases,

$$\text{Proportion of exposure} = \frac{84}{397}$$
$$= .212 \text{ or } 21.2 \text{ percent.}$$

The results reveal different exposure histories: The proportion among cases was higher than that among controls.

Similar examination of the data for nonsmokers shows that, by taking into consideration the numbers of cases and of controls, we have the following figures for employment:

(i) for the controls,

$$\text{Proportion of exposure} = \frac{35}{238}$$
$$= .147 \text{ or } 14.7 \text{ percent;}$$

(ii) for the cases,

$$\text{Proportion of exposure} = \frac{11}{61}$$
$$= .180 \text{ or } 18.0 \text{ percent.}$$

The results also reveal different exposure histories: The proportion among cases was higher than that among controls. The term "exposure" is used here to emphasize that employment in shipyards is a suspected "risk" factor.

The above analyses also show that the difference between proportions of exposure among smokers, that is,

$$21.2 - 14.3 = 6.9\%,$$

is different from the difference between proportions of exposure among nonsmokers, which is

$$18.0 - 14.7 = 3.3\%.$$

In other words, the possible effects of employment in shipyards (as a suspected risk factor) are different for smokers and nonsmokers. This difference of differences, if confirmed, is called a "three-term interaction" or an "effect modification," where smoking alters the effect of employment in shipyards as a risk for lung cancer. In some extreme examples, a pair of variables may even have their marginal association of different direction from their partial association (the association between them as seen at each and every level of a confounder or effect modifier). This interesting phenomenon is called *Simpson's paradox*, which further emphasizes the analysis complexity when we have data involving more than two variables.

1.2. A REVIEW OF LIKELIHOOD-BASED METHODS

Problems in biological and health sciences are formulated mathematically by considering the data that are to be used for making a decision as the observed values of a certain random variable X. The distribution of X is assumed to belong to a certain family of distributions specified by one or several parameters. The problem for decision makers is to decide on the basis of the data which members of the family could represent the distribution of X, that is, to predict or estimate the value of a parameter θ (or several parameters). The magnitude of a parameter often represents the effect of a risk or environmental factor and knowing its value, even approximately, would shed some light on the impact of such a factor. The likelihood function $L(x;\theta)$ for random sample $\{x_i\}$ of size n from the probability density function (pdf) $f(x;\theta)$ is

$$L(x;\theta) = \prod_{i=1}^{n} f(x_i;\theta).$$

The maximum likelihood estimator (MLE) of θ is the value $\hat{\theta}$ for which $L(x;\theta)$ is maximized. Calculus suggests setting the derivative of L with respect to θ equal to zero and solving the resulting equation. Since

$$\frac{dL}{d\theta} = L\frac{d\ln L}{d\theta},$$

$dL/d\theta = 0$ when and only when $d\ln L/d\theta = 0$ because L is never zero. Thus we can find the possible maximum of L by examining $\ln L$,

$$\ln L = \sum_{i=1}^{n} \ln f(x_i;\theta).$$

(It is often easier to deal mathematically with a sum than with a product.)

The maximum likelihood estimator (MLE) has a number of good properties which we will state without proofs; readers can skip this section without having any discontinuity.

(i) MLE is consistent.

(ii) If an efficient estimator exists, it is the MLE.

(iii) The MLE is asymptotically distributed as normal. The variance of this asymptotic distribution is given by

$$\mathrm{Var}(\hat{\theta}) = 1/E\left\{-\frac{d^2\ln L}{d\theta^2}\right\}$$

in which the denominator is sometimes estimated by $-d^2\ln L/d\theta^2$ evaluated at MLE $\hat{\theta}$.

It is noted that this is an asymptotic distribution, i.e., results are good for large samples only. If a closed-form solution does not exist, the iterative solution may be obtained by first solving for an additive correction

$$\Delta\hat{\theta} = -\left(\frac{d\ln L}{d\theta}\right) \Big/ \left(\frac{d^2\ln L}{d\theta^2}\right)$$

using numerical values of the derivatives. The iterative solution by this Newton–Raphson method would proceed as follows:

(i) Provide an initial value of $\hat{\theta}$, denoted by $\hat{\theta}^{(0)}$.

(ii) Solve for $\Delta\hat{\theta}$, evaluating the derivatives at $\hat{\theta}^{(0)}$.

(iii) Add $\Delta\hat{\theta}$ to the initial value to obtain a new value for $\hat{\theta}$, i.e.,

$$\hat{\theta}^{(1)} = \hat{\theta}^{(0)} + \Delta\hat{\theta}.$$

$\Delta\hat{\theta}$ is evaluated using $\hat{\theta}^{(0)}$.

(iv) Repeat steps (ii) and (iii) using $\hat{\theta}^{(1)}$ to obtain $\hat{\theta}^{(2)}$, etc. (and stop when results from successive steps are very close).

Estimate of variance is then given by

$$\widehat{\text{Var}}(\hat{\theta}) = 1 \bigg/ \left\{ -\frac{d^2 \ln L}{d\theta^2} \right\}$$

where the second derivative is evaluated using the value of the last iteration, i.e., $\hat{\theta}$, the MLE of θ.

For example, we have for a binomial distribution (more in Chapter 2)

$$L(x; p) = \binom{n}{x} p^x (1-p)^{n-x}$$

$$\ln L(x; p) = \ln \binom{n}{x} + x \ln p + (n-x) \ln(1-p).$$

From

$$\frac{d[\ln L(x; p)]}{dp} = \frac{x}{p} - \frac{n-x}{1-p}$$

we have

$$\hat{p} = x/n \quad \text{(the sample proportion).}$$

We also have

$$-\frac{d^2 \ln L}{dp^2} = \frac{x}{p^2} + \frac{n-x}{(1-p)^2}$$

leading to

$$E\left\{-\frac{d^2 \ln L}{dp^2}\right\} = \frac{np}{p^2} + \frac{n-np}{(1-p)^2}$$
$$= \frac{n}{p(1-p)}$$

or

$$\text{Var}(\hat{p}) = 1/E\left\{-\frac{d^2 \ln L}{dp^2}\right\}$$
$$= \frac{p(1-p)}{n}.$$

Consider the case of a two-parameter model. Let $L(x;\theta_1,\theta_2)$ be the likelihood function defined from a random sample $\{x_i\}$ by

$$L(x;\theta_1,\theta_2) = \prod_{i=1}^{n} f(x_i;\theta_1,\theta_2).$$

The MLEs $\hat{\theta}_1$ and $\hat{\theta}_2$ of θ_1 and θ_2 are the values for which $L(x;\theta_1,\theta_2)$ is maximized; these estimates are obtained by solving the following equations:

$$\frac{\partial \ln L}{\partial \theta_1} = 0$$
$$\frac{\partial \ln L}{\partial \theta_2} = 0.$$

If closed-form solutions do not exist, the iterative solutions to these equations may be obtained by solving the following,

$$\frac{\partial \ln L}{\partial \theta_1} = \left(-\frac{\partial^2 \ln L}{\partial \theta_1^2}\right)(\Delta\hat{\theta}_1) + \left(-\frac{\partial^2 \ln L}{\partial \theta_1 \partial \theta_2}\right)(\Delta\hat{\theta}_2)$$

$$\frac{\partial \ln L}{\partial \theta_2} = \left(-\frac{\partial^2 \ln L}{\partial \theta_1 \partial \theta_2}\right)(\Delta\hat{\theta}_1) + \left(-\frac{\partial^2 \ln L}{\partial \theta_2^2}\right)(\Delta\hat{\theta}_2),$$

where $\Delta\hat{\theta}_1$ and $\Delta\hat{\theta}_2$ are additive corrections. The iterative solution by the Newton–Raphson method would proceed as follows:

(i) Provide initial values of $\hat{\theta}_1$ and $\hat{\theta}_2$, denoted $\hat{\theta}_1^{(0)}$ and $\hat{\theta}_2^{(0)}$.

(ii) Solve for $\Delta\hat{\theta}_1$ and $\Delta\hat{\theta}_2$, evaluating the partial derivatives at the initial values $\hat{\theta}_1^{(0)}$ and $\hat{\theta}_2^{(0)}$.

(iii) Add $\Delta\hat{\theta}_1$ and $\Delta\hat{\theta}_2$ to the initial values $\hat{\theta}_1^{(0)}$ and $\hat{\theta}_2^{(0)}$ respectively to obtain new values of $\hat{\theta}_1$ and $\hat{\theta}_2$, called $\hat{\theta}_1^{(1)}$ and $\hat{\theta}_2^{(1)}$.

(iv) Repeat steps (ii) and (iii) using $\hat{\theta}_1^{(1)}$ and $\hat{\theta}_2^{(1)}$, etc. (and stop when results from successive steps are close).

The variance-covariance matrix of the estimates $\hat{\theta}_1, \hat{\theta}_2$ is given by I^{-1} and Fisher's information matrix I is defined as

$$I = \begin{bmatrix} E\left(-\frac{\partial^2 \ln L}{\partial \theta_1^2}\right) & E\left(-\frac{\partial^2 \ln L}{\partial \theta_1 \partial \theta_2}\right) \\ E\left(-\frac{\partial^2 \ln L}{\partial \theta_1 \partial \theta_2}\right) & E\left(-\frac{\partial^2 \ln L}{\partial \theta_2^2}\right) \end{bmatrix}.$$

In obtaining numerical estimates, expected values of the partial derivatives are replaced by numerical evaluations of those partial derivatives using values from the last iteration (solutions), i.e.,

$$\begin{bmatrix} \widehat{\mathrm{Var}}(\hat{\theta}_1) & \widehat{\mathrm{Cov}}(\hat{\theta}_1,\hat{\theta}_2) \\ \widehat{\mathrm{Cov}}(\hat{\theta}_1,\hat{\theta}_2) & \widehat{\mathrm{Var}}(\hat{\theta}_2) \end{bmatrix} = \begin{bmatrix} -\frac{\partial \ln L}{\partial \theta_1^2} & -\frac{\partial^2 \ln L}{\partial \theta_1 \partial \theta_2} \\ -\frac{\partial^2 \ln L}{\partial \theta_1 \partial \theta_2} & -\frac{\partial \ln L}{\partial \theta_1^2} \end{bmatrix}^{-1}.$$

Of course, the maximum likelihood procedure can be easily generalized to models with more than two parameters.

As an example of two-parameter models, consider a random sample of size n, $\{x_i\}$, from a normal distribution $N(\mu,\sigma^2)$. We have, with $\theta = \sigma^2$,

$$L(x;\mu,\theta) = \prod_{i=1}^{n} \frac{1}{\theta^{1/2}\sqrt{2\pi\pi}} \exp\left[-\frac{(x_i-\mu)^2}{2\theta}\right]$$

$$\ln L(x;\mu,\theta) = -\frac{n}{2}\ln\theta - \frac{n}{2}\ln(2\pi) - \frac{1}{2\theta}\sum_{i=1}^{n}(x_i-\mu)^2.$$

From

$$\frac{\partial \ln L}{\partial \mu} = \frac{1}{\theta}\sum_{i=1}^{n}(x_i-\mu)$$

$$\frac{\partial \ln L}{\partial \theta} = \frac{1}{2\theta}\left\{-n + \frac{1}{\theta}\sum_{i=1}^{n}(x_i-\mu)^2\right\}$$

we have

$$\hat{\mu} = \bar{x}$$

$$\hat{\sigma}^2 = \frac{1}{n}\sum_{i=1}^{n}(x_i-\bar{x})^2$$

$$= \frac{n-1}{n}s^2.$$

(Incidentally, this demonstrates that MLEs need not be unbiased—because s^2 is known to be an unbiased estimator of σ^2.)

The second-order derivatives

$$\frac{\partial^2 \ln L}{\partial \mu^2} = -n/\theta$$

$$\frac{\partial^2 \ln L}{\partial \mu \partial \theta} = -\frac{1}{\theta^2}\left\{\sum x_i - n\mu\right\}$$

$$\frac{\partial^2 \ln L}{\partial \theta^2} = \frac{n}{2\theta^2} - \frac{1}{\theta^3}\sum_{i=1}^{n}(x_i-\mu)^2$$

and their expected values lead to

$$\mathrm{Var}(\bar{x}) = \sigma^2/n$$

$$\mathrm{Var}(\hat{\sigma}^2) = 2\sigma^4/n$$

$$\mathrm{Cov}(\bar{x}, \hat{\sigma}^2) = 0.$$

A multiple regression model involves many parameters, the unknown regression coefficients β's; Once we have fit such a multiple regression model and obtained estimates for the various parameters of interest using the above method, we want to answer questions about the contributions of various factors to the prediction of the response variable. There are three types of such questions:

(i) An overall test: Taken collectively, does the entire set of explatory or independent variables contribute significantly to the prediction of the response? The null hypothesis for this test is

$$\mathcal{H}_0 : \beta_1 = \beta_2 = \cdots = \beta_k = 0.$$

Two likelihood-based statistics can be used to test this *global* null hypothesis; each has a symptotic chi-squared distribution with k degrees of freedom under $\mathcal{H}_0$:

(a) Likelihood Ratio Test:

$$\chi^2_{\mathrm{LR}} = 2[\ln L(\hat{\beta}) - \ln L(\mathbf{0})]$$

(b) Score Test:

$$\chi^2_S = \left[\frac{\delta \ln L(\mathbf{0})}{\delta\beta}\right]\left[-\frac{\delta^2 \ln L(\mathbf{0})}{\delta\beta^2}\right]^{-1}\left[\frac{\delta \ln L(\mathbf{0})}{\delta\beta}\right]$$

Both statistics are provided by most standard computer programs such as SAS and they are asymptotically equivalent yielding identical statistical decisions most of the times.

(ii) Test for the value of a single factor: Does the addition of one particular variable of interest add significantly to the prediction of response over and above that achieved by other independent variables? Let us assume that we now wish to test whether the addition of one particular independent variable of interest adds significantly to the prediction of the response over and above that achieved by other factors already present in the model. The null hypothesis for this test may stated as: "Factor X_i does not have any value added to the prediction of the response *given that other factors are already included in the model.*" In other words,

$$\mathcal{H}_0 : \beta_i = 0.$$

To test such a null hypothesis, one can perform a likelihood ratio chi-squared test, with 1 df, similar to that for the above global hypothesis:

$$\chi^2_{\text{LR}} = 2[\ln L(\hat{\beta};\text{all } X\text{'s}) - \ln L(\hat{\beta};\text{all other } X\text{'s with } X_i \text{ deleted})].$$

A much easier alternative method is using

$$z_i = \frac{\hat{\beta}_i}{\text{SE}(\hat{\beta}_i)},$$

where $\hat{\beta}_i$ is the corresponding estimated regression coefficient and $\text{SE}(\hat{\beta}_i)$ is the estimate of the standard error of $\hat{\beta}_i$, both of which are printed by standard computer packaged programs. In performing this test, we refer the value of the z statistic to percentiles of the standard normal distribution.

(iii) Test for contribution of a group of variables: Does the addition of a group of variables add significantly to the prediction of response over and above that achieved by other independent variables? This testing procedure addresses the more general problem of assessing the additional contribution of two or more factors to the prediction of the response over and above that made by other variables already in the regression model. In other words, the null hypothesis is of the form

$$\mathcal{H}_0 : \beta_1 = \beta_2 = \cdots = \beta_m = 0.$$

To test such a null hypothesis, one can perform a likelihood ratio chi-squared test, with m df,

$$\chi^2_{\text{LR}} = 2[\ln L(\hat{\beta};\text{all } X\text{'s}) - \ln L(\hat{\beta};\text{all other } X\text{'s with } X\text{'s under investigation deleted})].$$

This *multiple contribution* procedure is very useful for assessing the importance of potential explanatory variables. In particular, it is often used to test whether a similar group of variables, such as *demographic characteristics*, is important for the prediction of the response; these variables have some trait in common. Another application would be a collection of powers *and/or* product terms (referred to as interaction variables). It is often of interest to assess the interaction effects collectively before trying to consider individual interaction terms in a model as previously suggested. In fact, such use reduces the total number of tests to be performed and this, in turn, helps to provide better control of overall Type I error rate which may be inflated due to multiple testing.

(iv) In many applications, we wish to identify from many available factors a small subset of factors that relate significantly to the outcome (e.g., the disease under investigation). In that identification process, of course, we wish to avoid a large Type I (false positive) error. In a regression analysis, a Type I error corresponds to including a predictor that has no real relationship to the outcome; such an inclusion can greatly confuse the interpretation of the regression results. In a standard multiple regression analysis, this goal can be achieved by using a strategy that adds into or removes from a regression model one factor at a time according to a certain order of relative importance. Therefore, the two important steps are:

1. Specifying a criterion or criteria for selecting a model. The selection is often based on the likelihood ratio chi-squared statistic.
2. Specifying a strategy for applying the chosen criterion or criteria; such a strategy is concerned with whether any

particular variable should be added to a model or whether any variable should be deleted from a model at a particular stage of the process (stepwise regression). As computers became more accessible and more powerful, these practices became more popular.

1.3. ABOUT THIS BOOK

This book is intended to meet the need of practitioners and students in applied fields for a single, fairly thin volume covering major, updated methods in the analysis of categorical data. It is written for the training of beginning graduate students in biostatistics, epidemiology, and environmental health, as well as for biomedical research workers. As a book for biostatistics and statistics students, it is designed to offer some details for a better understanding of the various procedures as well as the relationships among different methods. However, the mathematics has been kept to an absolute minimum. As a book for students in applied fields and as a reference book for practicing biomedical research workers, *Applied Categorical Data Analysis* is application oriented. It introduces applied research areas, a large number of real-life examples, most of which are completely solved with samples of computer programs included.

The book is divided into seven chapters including this introductory chapter. Chapter 2 covers basic methods and applications of two-way contingency tables including etiologic fractions, the evaluation of ordinal risks, and the Mantel–Haenszel method. Chapter 3 is devoted to loglinear models; topics coved include the selection of the best model for three-way tables, and selection of a model for higher-dimensional tables, with or without the identification of a dependent variable. Chapter 4 is focused on logistic regression models, both binary and ordinal responses. Topics covered include stepwise procedure, measures of goodness-of-fit, and the use of logistic models for different designs. Chapter 5 covers similar topics as those in Chapters 2–4 but for matched designs, singly or multiply, including the conditional logistic regression model. Chapter 6 covers analytical methods for count data, including the Poisson regression model. Topics covered in this chapter include overdispersion and how to fit overdispersed models. The last chapter, Chapter 7, presents a brief introduction to survival analysis and Cox's regression model. In each chapter, numerous examples are provided for illustration.

CHAPTER 2

Two-Way Contingency Tables

This chapter presents basic inferential methods for categorical data, the analysis of two-way contingency tables.

Let X_1 and X_2 denote two categorical variables, X_1 having I levels and X_2 having J levels; there IJ combinations of classifications. We display the data in a rectangular table having I rows for the categories of X_1 and J columns for the categories of X_2. The IJ cells represent the IJ combinations of classifications; their probabilities are $\{\pi_{ij}\}$, where π_{ij} denotes the probability that (X_1, X_2) falls in the cell in row

i and column j. When the cells contain frequencies of outcomes, the table is called a contingency table or cross-classified table, also referred to as I-by-J or $I \times J$ table. We start with the most simple case, the two-by-two tables, and a special application in disease screening.

2.1. SCREENING TESTS

Important applications of proportions can be found in the study of screening tests or diagnostic procedures. Following these procedures, clinical observation or laboratory techniques, individuals are classified as healthy or as falling into one of a number of disease categories. Such tests are important in medicine and epidemiologic studies, and form the basis of early interventions. Almost all such tests are imperfect, in the sense that healthy individuals will occasionally be classified wrongly as being ill, while some individuals who are really ill may fail to be detected. Suppose that each individual in a large population can be classified as truly positive or negative for a particular disease; this true diagnosis may be based on more refined methods than are used in the test; or it may be based on evidence which emerges after passage of time, for instance, at autopsy.

The two proportions fundamental to evaluating diagnostic procedures are *sensitivity* and *specificity*. The sensitivity is the proportion of diseased individuals detected as positive by the test

$$\text{Sensitivity} = \frac{\text{Number of diseased individuals who screen positive}}{\text{Total number of diseased individuals}},$$

whereas the specificity is the proportion of healthy individuals detected as negative by the test

$$\text{Specificity} = \frac{\text{Number of healthy individuals who screen negative}}{\text{Total number of healthy individuals}}.$$

In statistical terms, these are conditional probabilities:

$$\text{Sensitivity} = \Pr(\text{Test} = + \mid \text{Disease} = +)$$

$$\text{Specificity} = \Pr(\text{Test} = - \mid \text{Disease} = -).$$

The complements of these quantities, Pr(Test = + | Disease = −) and Pr(Test = − | Disease = +), are the analogs of Type I and Type II errors in statistical tests of significance.

Example 2.1. A cytological test was undertaken to screen women for cervical cancer (May, 1974). Consider a group of 24,103 women consisting of 379 women whose cervices are abnormal (to an extent sufficient to justify concern with respect to possible cancer) and 23,724 women whose cervices are acceptably healthy. A test was applied and results are tabulated in the table below:

	Test Result		
Disease	Negative (−)	Positive (+)	Total
Negative (−)	23,362	362	23,724
Positive (+)	225	154	379

The results

$$\text{Sensitivity} = \frac{154}{379}$$

$$= .406 \text{ or } 40.6 \text{ percent}$$

$$\text{Specificity} = \frac{23{,}362}{23{,}724}$$

$$= .985 \text{ or } 98.5 \text{ percent}$$

show that the above test is highly specific (98.5%) but not very sensitive (40.6%); there were more than half (59.4%) false negatives. The implications of the use of this test are

(i) if a woman without cervical cancer is tested, the result would almost surely be negative, *but*

(ii) if a woman with cervical cancer is tested the chance is that the disease would go undetected because 59.4% of these cases would lead to false negatives.

In addition to the sensitivity, Pr(Test = + | Disease = +), and specificity, Pr(Test = − | Disease = −), there are two other important proportions:

$$\text{Positive Predictivity} = \Pr(\text{Disease} = + \mid \text{Test} = +)$$

$$\text{Negative Predictivity} = \Pr(\text{Disease} = - \mid \text{Test} = -).$$

With positive predictivity (or positive predictive value), the question is, given that the test result suggests cancer, what is the probability that, in fact, cancer is present? Rationales for these predictivities are that a test passes through several stages. Initially, the original test idea occurs to some researcher. It then must go through a developmental stage. This may have many aspects (in biochemistry, microbiology, etc.) one of which is in biostatistics (viz. trying the test out on a pilot population). From this developmental stage, efficiency of the test is characterized by the sensitivity and specificity. An efficient test will then go through an applicational stage with an actual application of the screening test to a target population; and here we are concerned with its predictive values. The following simple example shows that, unlike sensitivity and specificity, the positive and negative predictive values depend not only on the efficiency of the test but also on the disease prevalence of the target population:

Population A

		Test Result +	Test Result −
Disease	+	45,000	5,000
	−	5,000	45,000

Population B

		Test Result +	Test Result −
Disease	+	9,000	1,000
	−	9,000	81,000

In both cases, the test is 90 percent sensitive and 90 percent specific. However,

1. Population A has a prevalence of 50%, leading to a positive predictive value of 90%.
2. Population B has a prevalence of 10%, leading to a positive predictive value of 50%.

The conclusion is clear: If a test—even a highly sensitive and highly specific one—is applied to a target population in which the disease prevalence is low (for example, population screening for a rare disease), the positive predictive value is low. (How does this relate to an important public policy: Should we conduct random testing for AIDS?)

In the actual application of a screening test to a target population (the applicational stage), data on the disease status of individuals are not available (otherwise, screening would not be needed). However, disease prevalences are often available from national agencies and health surveys. Predictive values are then calculated from

Positive predictivity

$$= \frac{(\text{Prevalence})(\text{Sensitivity})}{(\text{Prevalence})(\text{Sensitivity}) + (1 - \text{Prevalence})(1 - \text{Specificity})}$$

and

Negative predictivity

$$= \frac{(1 - \text{Prevalence})(\text{Specificity})}{(1 - \text{Prevalence})(\text{Specificity}) + (\text{Prevalence})(1 - \text{Sensitivity})}.$$

These formulas, from the *Bayes theorem*, allow us to calculate the predictivities without having data from the application stage. All we need are the disease prevalence (obtainable from federal health agencies) and sensitivity and specificity; these were obtained after the developmental stage.

Sensitivity and specificity are sample proportions; besides the point estimates, we can calculte the confidence intervals for these parameters using standard statistical methods, e.g., point estimate ± 1.96 (standard error). Positive predictive value and negative predictive value are derived from sensitivity, specificity, and disease prevalence; their standard errors can be obtained using the delta method for errors propagation (disease prevalence, which is usually obtained from a very large national survey, is considered as fixed).

2.2. SOME SAMPLING MODELS FOR CATEGORICAL DATA

In this section we will describe briefly several discrete distributions which are often chosen as models for categorical data. Since we expect this section to be used only as brief reference, we assume that its readers have some familiarity with basic statistical terminology.

2.2.1. The Binomial and Multinomial Distributions

The binomial model applies when each trial of an experiment has two possible outcomes (often referred to as "failure" and "success," or "negative" and "positive"; one has a success when the primary outcome is observed). Let the probabilities of failure and success be, respectively, $1-\pi$ and π, and we "code" these two outcomes as 0 (zero successes) and 1 (one success). The experiment consists of n repeated trials satisfying these assumptions:

(i) The n trials are all independent.
(ii) The parameter π is the same for each trial.

The model is concerned with the total number of successes in n trials as a random variable, denoted by X. Then it can be seen that

$$\Pr(X=x) = \binom{n}{x}\pi^x(1-\pi)^{n-x} \qquad \text{for} \quad x=0,1,2,\ldots,n$$
$$= b(x;n,\pi)$$

where $\binom{n}{x}$ is the number of combinations of x objects selected from a set of n objects,

$$\binom{n}{x} = \frac{n!}{x!(n-x)!}.$$

The above formula for $\Pr(X=x)$ is intuitively obvious because there is one factor π for each of the x successes, one factor $(1-\pi)$ for each of the $(n-x)$ failures; these factors are all multiplied together by the virtue of the assumption that the n trials are independent. In addition, the product of n factors applies to any sequence of n trials in which there are x successes and $(n-x)$ failures in any order; the number of such sequences is $\binom{n}{x}$. We refer to the above probability function as the *binomial distribution* and it is specified by two parameters n and π, the number of trials and the probability of success for each trial.

The mean and variance of the binomial distribution are:

$$\mu = n\pi$$

$$\sigma^2 = n\pi(1-\pi).$$

(In some real-life problems, the true variance may be larger than $n\pi(1-\pi)$, a phenomenon known as overdispersion. We will deal with this within the context of logistic regression analysis in Chapter 4). If X is a binomial random variable with parameter n and π, then the maximum likelihood estimate of π is the sample proportion p whose standard error is given by

$$\mathrm{SE}(p) = \sqrt{\frac{p(1-p)}{n}},$$

from which we can form confidence intervals for π.

An obvious generalization of the binomial is the multinomial distribution, which arises when each trial has more than two possible outcomes. Let k be the number of mutually exclusive outcomes whose respective probabilities are $\pi_1, \pi_2, \ldots, \pi_k$ and $\sum_{i=1}^{k} \pi_i = 1$. Let $X_1, \ldots, X_k$ be the k random variables defined so that X_i is the number of times (out of a total of n trials) that the ith outcome occurs; then the joint probability function of the X's is given by

$$f(x_1, x_2, \ldots, x_k) = \frac{n!}{x_1! x_2! \cdots x_k!} \pi_1^{x_1}, \pi_2^{x_2} \cdots \pi_k^{x_k},$$

where $x_i = 1, 2, \ldots, n$ and $\sum_{i=1}^{k} x_i = n$.

Using this multinomial pdf, it can be shown that the means, variances, and covariances are given by

$$\begin{aligned}\mu_i &= E(X_i)\\ &= n\pi_i\\ \sigma_i^2 &= \mathrm{Var}(X_i)\\ &= n\pi_i(1-\pi_i)\end{aligned}$$

and, for $i \neq j$,

$$\mathrm{Cov}(X_i, X_j) = -n\pi_i\pi_j$$

(i.e., X_i and X_j are not independent).

Example 2.2. The unknown number X of cells in a solution is estimated using the following experiment:

Step 1: A number Y of "beads" was added to the solution.
Step 2: n units from the solution were sampled which was found to contain x cells and $(n-x)$ beads.

If the solution was mixed well, we have

$$\frac{X}{Y} = \frac{x}{n-x},$$

from which we obtain

$$\begin{aligned}\hat{X} &= Y \cdot \frac{x}{n-x}\\ &= Y \cdot \frac{p}{1-p},\end{aligned}$$

where p is the sample proportion:

$$p = \frac{x}{n}.$$

It can be shown that the cells counting experiment is "optimal", i.e., when the coefficient of variation of $\hat{X}$ is minimized, if $p = 1/2$.

2.2.2. The Hypergeometric Distributions

Sampling (selecting objects) may be done with or without replacement. To develop a formula or model analogous to that of the binomial probability which applies to sampling without replacement, in which case successive trials are not independent, let us consider a set or population having $(a+b)$ elements of which a are labeled "success" and b are labeled "failure." Suppose a sample of n objects are taken from this population and let X denote the number of successes in this sample. Then it can be shown that

$$\Pr(X = x) = \frac{\binom{a}{x}\binom{b}{n-x}}{\binom{a+b}{n}} \qquad \text{for} \quad x = 0, 1, 2, \ldots, n$$

$$= h(x; n, a, b),$$

subject to the restrictions that x cannot exceed a and $n - x$ cannot exceed b. We refer to this probability function as the *hypergeometric distribution* $H(n, a, b)$ and it is specified by three parameters n, a, and b.

The mean and variance of the hypergeometric distribution are

$$\mu = na/(a+b)$$

$$\sigma^2 = \frac{nab(a+b-n)}{(a+b)^2(a+b-1)}.$$

In the case of a large population and n small as compared to a and b, these results are similar to those of the binomial with $\pi = a/(a+b)$. In the general case, Wise (1954) gave the following approximation for cumulative hypergeometric probabilities:

$$\sum_{x=0}^{k} h(x; n, a, b) = \sum_{x=0}^{k} b(x; n, \pi)$$

in which π is given by

$$\pi = (a - k/2)/(a + b - n/2 + .5).$$

For example, let $a = 6$, $b = 12$, $n = 5$ and $k = 1$, we have .439 for the left-hand side and .440 for the right-hand side.

An obvious generalization of the hypergeometric is the multivariate hypergeometric distribution, which arises when the population has more than two types of individuals. Consider a population of N subjects, of which N_i are of type i, $1 \le i \le k$. If a sample of size n is taken without replacement and let $X_1, \ldots, X_k$ be the k random variables defined so that X_i is the number of subjects (out of a total of n subjects in the sample) that belong to type i, then the joint probability function of the X's is given by

$$f(x_1, x_2, \ldots, x_k) = \frac{\prod_{i=1}^{k} \binom{N_i}{x_i}}{\binom{N}{n}},$$

where $x_i = 1, 2, \ldots, N_i$ and $\sum_{i=1}^{k} x_i = n$.

Using this multivariate pdf, it can be shown that the means, variances, and covariances are given by

$$\begin{aligned} \mu_i &= E(X_i) \\ &= \frac{nN_i}{N} \\ \sigma_i^2 &= \mathrm{Var}(X_i) \\ &= n\left\{\frac{N-n}{N-1}\right\}\left\{\frac{N_i}{N}\right\}\left\{1 - \frac{N_i}{N}\right\} \end{aligned}$$

and for $i \neq j$,

$$\mathrm{Cov}(X_i, X_j) = -\frac{nN_iN_j}{N^2}\left\{\frac{N-n}{N-1}\right\}$$

(i.e., X_i and X_j are not independent, similar to the case of the multinomial).

For the hypergeometric distribution and its multivariate generalization, the multivariate hypergeometric distribution, the parameter estimation procedures are less often needed; these models are often used when conditional arguments are being used or conditional tests are being considered (see section 2.3.6).

2.3. INFERENCES FOR TWO-BY-TWO TABLES

Data for two-by-two tables may be obtained in a different way. In the following two examples, the first is the result of a case-control study (two independent binomial samples) and the second comes from a cross-sectional survey (one multinomial sample with four categories).

Example 2.3. the role of smoking in the etiology of pancreatitis has been recognized for many years. In order to provide estimates of the quantitative significance of these factors, a hospital-based study was carried out in eastern Massachusetts and Rhode Island between 1975 and 1979 (Yen et al., 1982). Ninety-eight patients who had a hospital discharge diagnosis of pancreatitis were included in this unmatched case-control study. The control group consisted of 451 patients admitted for diseases other than those of the pancreas and biliary tract. Risk factor information was obtained from a standardized interview with each subject, conducted by a trained interviewer.

The following are some data for the males:

Use of Cigarettes	Cases	Controls
Current Smokers	38	81
Never or Ex-smokers	15	136
Total	53	217

Example 2.4. In 1979 the U.S. Veterans Administration conducted a health survey of 11,230 veterans (True et al., 1988). The advantages of this survey are that it includes a large random sample with a high interview response rate and it was done before the

recent public controversy surrounding the issue of the health effects of possible exposure to Agent Orange. The following are data relating Vietnam service to having sleep problems among the 1787 veterans who entered the military service between 1965 and 1975.

	Service in Vietnam		
Sleep Problems	Yes	No	Total
Yes	173	160	333
No	599	851	1450
Total	772	1011	1783

2.3.1. Comparison of Two Proportions

Perhaps the most common problem involving categorical data is the comparison of two proportions calculated from two independent samples such as those in Example 2.3.

In this type of problem, we have two independent samples of binary data (n_1, x_1) and (n_2, x_2) where the n's are adequately large sample sizes that may or may not be equal, the x's are the numbers of "positive" outcomes in the two samples, and we consider the null hypothesis

$$\mathcal{H}_0 : \pi_1 = \pi_2$$

expressing the equality of the two population proportions.

To perform a test of significance for $\mathcal{H}_0$, we proceed with the following steps:

(i) Decide whether a one-tailed test, say

$$\mathcal{H}_A : \pi_2 > \pi_1,$$

or a two-tailed test,

$$\mathcal{H}_A : \pi_1 \neq \pi_2,$$

is appropriate.

(ii) Choose a significance level α, a common choice being .05.

(iii) Calculate the z-score

$$z = \frac{p_2 - p_1}{\sqrt{p(1-p)\left(\frac{1}{n_1} + \frac{1}{n_2}\right)}}$$

where p is the "pooled proportion" defined by

$$p = \frac{x_1 + x_2}{n_1 + n_2}.$$

(iv) Refer to the table for standard normal distribution, which can be found in any elementary book, for selecting a cut point. For example, if the choice of α is .05 then the rejection region is determined by the following:

- for the one-tailed alternative $\mathcal{H}_A : \pi_2 > \pi_1$, $z \geq 1.65$.
- for the one-tailed alternative $\mathcal{H}_A : \pi_2 < \pi_1$, $z \leq -1.65$.
- for the two-tailed test $\mathcal{H}_A : \pi_1 \neq \pi_2$, $z \leq -1.96$ or $z \geq 1.96$.

In the context of a case-control study, such as Example 2.3, n_1 is the number of controls, n_2 is the number of cases, and the x's are the numbers of exposed subjects in the two samples.

Example 2.5. In Example 1.1, a case-control study was conducted to identify reasons for the exceptionally high rate of lung cancer among male residents of coastal Georgia (Blot et al., 1978). The primary risk factor under investigation was employment in shipyards during World War II, and the following table provides data for nonsmokers:

Shipbuilding	Cases	Controls
Yes	11	35
No	50	203
Total	61	238

We have for the cases

$$\begin{aligned} p_2 &= 11/61 \\ &= .180 \end{aligned}$$

and for the controls

$$\begin{aligned} p_1 &= 35/238 \\ &= .147. \end{aligned}$$

An application of the procedure yields a pooled proportion of

$$\begin{aligned} p &= \frac{11+35}{61+238} \\ &= .154 \end{aligned}$$

leading to

$$\begin{aligned} z &= \frac{.180-.147}{\sqrt{(.154)(.846)\left(\frac{1}{61}+\frac{1}{238}\right)}} \\ &= .64. \end{aligned}$$

It can be seen that the rate of employment for the cases (18.0%) was higher than that for the controls (14.7%) but the difference is not statistically significant at the .05 level ($z = .64 < 1.65$).

Example 2.6. Refering to the pancreatitis data of Example 2.3 with currently smoking being the exposure, we have for the cases

$$\begin{aligned} p_2 &= 38/53 \\ &= .717 \end{aligned}$$

and for the controls

$$\begin{aligned} p_1 &= 81/217 \\ &= .373. \end{aligned}$$

An application of the procedure yields a pooled proportion of

$$p = \frac{38+81}{53+217} = .441$$

leading to

$$z = \frac{.717-.373}{\sqrt{(.441)(.559)\left(\frac{1}{53}+\frac{1}{217}\right)}} = 4.52.$$

It can be seen that the proportion of smokers among the cases (71.7%) was higher than that for the controls (37.7%) and the difference is highly statistically significant ($p < .001$).

2.3.2. Tests for Independence

When we examine the Veterans Administration data in Example 2.4, the method of section 2.3.1 seems not applicable because we do not have two independent binomial samples. What we have is a multinomial sample with probabilities $\{\pi_{ij}\}$. For contingency tables, the maximum likelihood (ML) estimates of cell probabilities are the sample cell proportion, and the ML estimates of marginal probabilities are sample marginal proportions. When two categorical variables forming the two-way table are independent, all $\pi_{ij} = \pi_{i+}\pi_{+j}$. The ML estimate of π_{ij} under this condition is:

$$\widehat{\pi_{ij}} = \widehat{\pi_{i+}}\widehat{\pi_{+j}} = p_{i+}p_{+j} = \frac{x_{i+}x_{+j}}{n^2}.$$

For a multinomial sample of size n over $2 \times 2 = 4$ cells, an individual cell count x_{ij} has the binomial distribution with size n and probability

π_{ij}. The mean of this binomial is $m_{ij} = n\pi_{ij}$, which has ML estimate, under the assumption of independence:

$$\begin{aligned}\widehat{m_{ij}} &= n\widehat{\pi_{ij}} \\ &= \frac{x_{i+}x_{+j}}{n} \\ &= \frac{\text{(row total)(column total)}}{\text{sample size}}.\end{aligned}$$

The $\{\widehat{m_{ij}}\}$ are called estimated expected frequencies, the frequencies we expect to have under the null hypothesis of independence. They have the same marginal totals as do the observed data.

We use the $\{\widehat{m_{ij}}\}$ in tests of independence through the Pearson's chi-square statistic,

$$X^2 = \sum_{i,j} \frac{(x_{ij} - \widehat{m_{ij}})^2}{\widehat{m_{ij}}},$$

or the likelihood ratio chi-square statistic:

$$G^2 = 2\sum_{i,j} x_{ij} \log \frac{x_{ij}}{\widehat{m_{ij}}}.$$

For large samples, both X^2 and G^2 have approximately a chi-squared distribution with 1 degree of freedom under the null hypothesis of independence; greater values lead to a rejection of H_0. The Pearson's chi-quare statistic is more often used; as applied to the studies with two binomial samples, it is identical to the z-test of section 2.2.1.

Example 2.7. Refering to the Veterans Administration data of Example 2.4, we have

$$\begin{aligned}\widehat{m_{11}} &= \frac{(333)(772)}{1783} \\ &= 144.18\end{aligned}$$

$$\widehat{m_{12}} = 333 - 144.18$$

$$= 188.82$$

$$\widehat{m_{21}} = 772 - 144.18$$

$$= 627.82$$

$$\widehat{m_{22}} = 1011 - 188.82$$

$$= 822.18$$

leading to

$$X^2 = \frac{(173 - 144.18)^2}{144.18} + \frac{(160 - 188.82)^2}{188.82}$$

$$+ \frac{(599 - 627.82)^2}{627.82} + \frac{(851 - 822.18)^2}{822.18}$$

or $X^2 = 12.49$; and

$$G^2 = 2\left\{173 \log \frac{173}{144.18} + 160 \log \frac{160}{188.82}\right.$$

$$\left. + 599 \log \frac{599}{627.82} + 851 \log \frac{851}{822.18}\right\}$$

or $G^2 = 12.40$. These statistics, both with 1 df, indicate a significant correlation ($p < .001$) relating Vietnam service to having sleep problems among the veterans.

Statistical decisions based on the Pearson's chi-square statistic (and the likelihood ratio chi-square statistic for the same purpose) make use of the percentiles of the chi-square distribution. Since chi-square is a continuous distribution and categorical data are discrete, some statisticians use a version of the Pearson's statistic with a *continuity correction*, called Yates corrected chi-square test, which can be expressed as

$$X_c^2 = \sum_{i,j} \frac{(|x_{ij} - \widehat{m_{ij}}| - 0.5)^2}{\widehat{m_{ij}}}.$$

Statisticians still disagree about whether or not a continuity correction is needed (Conover, 1974). Generally, the corrected version is more conservative and more widely used in applied literature.

2.3.3. Fisher's Exact Test

Even with a continuity correction, the *goodness-of-fit* test statistics such as Pearson's X^2 and the likelihood ratio G^2 are not suitable when the sample is small. Generally, statisticians suggest using them only if no expected frequency in the table is less than 5. For studies with small samples, we will introduce a method known as *Fisher's exact test*. For tables in which the use of the chi-square test X^2 is appropriate, the two tests give very similar results.

Our purpose is to find the exact significance level associated with an observed table. The central idea is to enumerate all possible outcomes consistent with a given set of marginal totals and add up the probabilities of those tables more extreme than the one observed. Conditional on the margins, a 2×2 table is a one-dimensional random variable having a hypergeometric distribution so the exact test is relatively easy to implement. The probability of observing a table with cells a, b, c, and d (with total n) is

$$\Pr(a,b,c,d) = \frac{(a+b)!(c+d)!(a+c)!(b+d)!}{n!a!b!c!d!}.$$

The process for doing hand calculations would be as follows:

(i) Rearrange the rows and columns of the observed table so that the smaller total is in the first row and the smaller column total is in the first column.

(ii) Start with the table having 0 in the $(1,1)$ cell (top-left cell). The other cells in this table are determined automatically from the fixed row margins and column margins.

(iii) Construct the next table by increasing the $(1,1)$ cell from 0 to 1 and decreasing all other cells accordingly.

(iv) Continue to increase the $(1,1)$ cell by 1 until one of the other cells becomes 0. At that point we have enumerated all possible tables.

(v) Calculate and add up the probabilities of those tables with cell $(1,1)$ having values from 0 to the observed frequency (left tail for a one-tailed test); double the smaller tail for a two-tailed test.

In practice, the calculations are often tedious and should be left to a computer program to implement.

Example 2.8. A study on deaths of men aged over 50 yields the following data (numbers in parentheses are expected frequencies):

	Type of Diet		
Cause of Death	High Salt	Low Salt	Total
Non-CVD	2(2.92)	23(22.08)	25
CVD	5(4.08)	30(30.92)	35
Total	7	53	60

An application of the Fisher's exact test yields a one-tailed p-value of .375 or a two-tailed p-value of .688; we cannot say, on the basis of this limited amount of data, that there is a significant association between salt intake and cause of death even though the proportions of CVD-deaths are different (71.4% vs. 56.6%). For implementing hand calculations, we would focus on the tables where cell $(1,1)$ equals 0, 1, and 2 (observed value; the probabilities for these tables are 0.017, 0.105, and 0.252, respectively).

Note: An SAS program would include these instructions:

```
DATA;
DO CVD=1 TO 2;
DO DIET=1 TO 2;
INPUT COUNT @@; OUTPUT;
```

```
END;
END;
CARDS;
2 23
5 30;
PROC FREQ;
WEIGHT COUNT;
TABLES CVD*DIET;
```

The output also includes Pearson's test ($X^2 = 0.559$; $p = 0.455$) and the likelihood ratio test ($G^2 = 0.581$; $p = 0.446$) as well.

2.3.4. Relative Risk and Odds Ratio

One of the most often used ratios in epidemiological studies is the *relative risk* or *risks ratio*, a concept for the comparison of two groups or sub-populations with respect to a certain unwanted event (disease or death). The traditional method of expressing it in prospective studies is simply the ratio of the incidence rates:

$$\text{Relative Risk} = \frac{\text{Disease incidence in group 1}}{\text{Disease incidence in group 2}}.$$

We can also use ratio of disease prevalences as well as follow-up death rates. Usually, group 2 is under standard conditions—such as nonexposure to a certain risk factor—against which group 1 (exposed) is measured. For example, if group 1 consists of smokers and group 2 nonsmokers, then we have a relative risk due to smoking. The relative risk is an important index in epidemiological studies because in such studies it is often useful to measure the increased risk (if any) of incurring a particular disease if a certain factor is present. In cohort studies such an index is readily obtained by observing the experience of groups of subjects with and without the factor as shown above. In a case-control study or cross-sectional survey, the data do not present an immediate answer to this type of question, and we now consider how to obtain a useful solution.

Suppose that each subject in a large study, at a particular time, is classified as positive or negative according to some risk factor, and as having or not having a certain disease under investigation. For any such categorization the population may be enumerated in a

2×2 table, as follows:

	Disease Classification		
Factor	+	−	Total
+	A	B	$A+B$
−	C	D	$C+D$
Total	$A+C$	$B+D$	N

The entries A, B, C, and D in the table are sizes of the four combinations of disease presence and factor presence and N is the total population size. The relative risk (RR) is

$$\begin{aligned} \text{RR} &= \frac{A}{A+B} \div \frac{C}{C+D} \\ &= \frac{A(C+D)}{C(A+B)}. \end{aligned}$$

In many situations, the number of subjects classified as disease positive is small compared to the number classified as disease negative, that is,

$$C + D \cong D$$
$$A + B \cong B,$$

and, therefore, the relative risk can be approximated as follows:

$$\begin{aligned} \text{RR} &\cong \frac{AD}{BC} \\ &= \frac{A/B}{C/D} \\ &= \frac{A/C}{B/D} \end{aligned}$$

(the slash denotes division, and $\cong$ means "almost equal to"). The resulting ratio, AD/BC, is an approximate relative risk, but it is often referred to as "odds ratio" because

(i) A/B and C/D can be thought of as odds in favor of having disease from groups with or without the factor;
(ii) A/C and B/D can be thought of as odds in favor of exposure to the factors from groups with or without the disease. The two odds can be easily estimated using case-control data, by using sample frequencies.

For the many diseases that are rare, the terms "relative risk" and "odds ratio" are used interchangeably. The relative risk is an important epidemiological index used to measure seriousness, or the magnitude of the harmful effect of suspected risk factors. For example, if we have

$$\mathrm{RR} = 3.0$$

we can say that the exposed individuals have a risk of contracting the disease which is three times the risk of unexposed individuals. A perfect 1.0 indicates no effect and beneficial factors result in relative risk values which are smaller than 1.0. From data obtained by a case-control (retrospective) or cross-sectional study it is impossible to calculate the relative risk that we want, but if it is reasonable to assume that the disease is rare (prevalence is less than .05, say) then we can calculate the odds ratio as a "stepping stone" and use it as an approximate relative risk (we use the notation $\cong$ for this purpose). In these cases, we interpret the calculated odds ratio just as we would do with the relative risk.

Data from a case-control study, for example, may be summarized in a 2×2 table:

	Exposed	Unexposed
Diseased	a	b
Disease-Free	c	d

The observed odds ratio (OR) is

$$\widehat{\mathrm{OR}} = \frac{ad}{bc}.$$

Confidence intervals are derived from the normal approximation to the sampling distribution of ln(OR) with

$$\widehat{\text{Variance}}[\ln(\widehat{\text{OR}})] \cong \frac{1}{a} + \frac{1}{b} + \frac{1}{c} + \frac{1}{d}.$$

Consequently, an approximate 95 percent confidence interval for odds ratio is given by *exponentiating* the two numbers

$$\ln \frac{ad}{bc} \pm 1.96\sqrt{\frac{1}{a} + \frac{1}{b} + \frac{1}{c} + \frac{1}{d}}.$$

(ln is logarithm to base e or natural logarithm.)

Example 2.9. Referring to the pancreatitis data of Example 2.3, we have

$$\widehat{\text{OR}} = \frac{(38)(136)}{(15)(81)}$$

$$= 4.25,$$

and a 95 percent confidence interval for the population odds ratio is

$$\text{Exp}\left[\ln 4.25 \pm 1.96\sqrt{\frac{1}{38} + \frac{1}{15} + \frac{1}{81} + \frac{1}{136}}\right] = (3.04, 5.95).$$

In other words, current smokers have the risk of contracting pancreatitis increased by 3.04 times to 5.95 times as compared to others not currently smoking.

2.3.5. Etiologic Fraction

The *etiologic fraction* or *population attributable risk* λ is a measure of the impact of an exposure *on the population* (whereas the *relative risk* measures the impact of the exposure *on the exposed subpopulation*). The etiologic fraction is defined as the proportion of disease cases *attributable* to the risk factor. For example, we are interested in the proportion of lung cancer cases attributable to smoking. At a given time, let N be the population size and N_1 be the number of cases. If p_1 and p_0 are the disease rates of the exposed and nonexposed subpopulations, respectively, then

$$\lambda = \frac{(N_1 - Np_0)}{N_1},$$

because if the exposure has no effect then the expected number of cases would be Np_0. Since

$$N_1 = N\{p_e p_1 + (1 - p_e)p_0\}$$

where p_e is the population exposure rate (e.g., percent of smokers in the target population), we have:

$$\lambda = \frac{p_e(\text{RR} - 1)}{1 + p_e(\text{RR} - 1)}$$

with RR being the relative risk associated with the exposure. This result shows that the attributable risk, or etiologic fraction, depends on the effect of the exposure, through RR, as well as the exposure rate, p_e:

(i) It is an increasing function of RR, $\lambda = 0$ if RR $= 1$, i.e., no *excess* risk.

(ii) It is also an increasing function of p_e, which makes it an important parameter for public health intervention policy; $\lambda = 0$ if $p_e = 0$.

However, the above formula is often *not* useful in practice because the population exposure rate, p_e, is unknown and not estimable accurately from case-control studies (we can estimate it using the controls, but it is only an approximation). Using the Bayes theorem, we can express the etiologic fraction in a different but equivalent form,

$$\lambda = p_{1e}\left(1 - \frac{1}{\text{RR}}\right),$$

where p_{1e} is the exposure rate of the subpopulation of cases which can be estimated using the sample of cases.

When data from a case-control study are presented as

Exposure	Cases	Controls
+	a	b
−	c	d
Sample Size	n_1	n_0

then we have

$$\hat{\lambda} = \frac{a}{n_1}\left\{1 - \frac{bc}{ad}\right\} \quad \text{or}$$

$$\hat{\lambda} = 1 - \frac{cn_0}{dn_1}$$

and a 95 percent confidence interval for λ is given by

$$1 - (1 - \hat{\lambda})\exp\left\{\pm 1.96\sqrt{\frac{a}{cn_1} + \frac{b}{dn_0}}\right\}.$$

Example 2.10. A case-control study was conducted in Auckland, New Zealand to investigate the effects of alcohol consumption on both nonfatal myocardial infarction and coronary death in the 24 hours after drinking, among regular drinkers (Jackson et al., 1992). The following table shows coronary death data for men:

Drink in the Last 24 Hours	Cases	Controls
Yes	69	159
No	103	135
Sample size	172	294

Here we have $n_1 = 172$ ($a = 69$, $c = 103$) and $n_0 = 294$ ($b = 159$, $d = 135$), leading to a point estimate of $\hat{\lambda} = -0.304$ and a 95 percent confidence interval: $(-0.552, -0.096)$. The results indicate a protective effect, that drinking reduces coronary deaths in the population of male drinkers in New Zealand between 9.6 percent to 55.2 percent.

2.3.6. Crossover Designs

The two-period crossover design is often used in clinical trials in order to improve the sensitivity of the trial by eliminating the individual patient effects. For two experimental treatments A and B, a set of N patients is randomly subdivided into n_A patients assigned to receive treatment sequence (A, B) and n_B assigned to (B, A), $N = n_A + n_B$ (usually, $n_A = n_B = N/2$). For the case of a binary response, coded 0/1 for negative/positive responses, Gart (1969) proposed the following so-called logistic response probabilities:

Sequence (A, B)

Treatment A:

$$\Pr(X_{i/A} = 1) = \exp(\beta_i + \lambda + \tau)/[1 + \exp(\beta_i + \lambda + \tau)]$$

Treatment B:

$$\Pr(X_{i/B} = 1) = \exp(\beta_i - \lambda - \tau)/[1 + \exp(\beta_i - \lambda - \tau)]$$

for $i = 1, 2, \ldots, n_A$.

Sequence (B, A)

Treatment A:

$$\Pr(X_{j/A} = 1) = \exp(\beta_j - \lambda + \tau)/[1 + \exp(\beta_j - \lambda + \tau)]$$

Treatment B:

$$\Pr(X_{j/B} = 1) = \exp(\beta_j + \lambda - \tau)/[1 + \exp(\beta_j + \lambda - \tau)]$$

for $j = 1, 2, \ldots, n_B$ in which the β's are the individual patient effects and λ, τ are the order and treatment effects, respectively.

Under this response model, Gart showed that optimum inferences about λ and τ, regarding the β's as nuisance parameters, are based on those patients with unlike responses in consecutive periods, i.e., those patients with outcomes (1,0) or (0,1) (a conditional approach).

Among these patients with unlike outcomes, let

Sequence (A, B)

$$y_A = \text{number of patients with outcome } (1,0)$$
$$y_B = \text{number of patients with outcome } (0,1)$$
$$n = y_A + y_B \leq n_A$$

Sequence (B, A)

$$y'_A = \text{number of patients with outcome } (0,1)$$
$$y'_B = \text{number of patients with outcome } (1,0)$$
$$n' = y'_A + y'_B \leq n_B$$

Then it can be shown that

(i) With n and n' being fixed, y_A and y'_A have the binomial distributions $B(n, \pi)$ and $B(n', \pi')$, respectively, where the parameters π and π' are given by

$$\pi = 1/\{1 + \exp[-2(\lambda + \tau)]\}$$

and

$$\pi' = 1/\{1 + \exp[2(\lambda - \tau)]\}.$$

(ii) With data presented as in the following 2×2 table,

Positive Responses Coming From:	Treatment Sequence (A, B)	(B, A)	Total
First period	y_a	y'_b	$y_a + y'_b$
Second period	y_b	y'_a	$y_b + y'_a$
Total	$n = y_a + y_b$	$n' = y'_b + y'_a$	$n + n'$

it can be seen that y_a has the hypergeometric distribution $H(y_a + y'_b, n, n')$ under the null hypothesis

$$\mathcal{H}_\tau : \tau = 0$$

of no treatment effects and conditional on the marginal totals. This can be tested using Pearson's chi-square test or Fisher's exact test.

(iii) With data presented as in the following 2×2 table,

Positive Responses Coming From:	Treatment Sequence (A, B)	(B, A)	Total
Treatment A	y_a	y'_a	$y_a + y'_a$
Treatment B	y_b	y'_b	$y_b + y'_b$
Total	$n = y_a + y_b$	$n' = y'_a + y'_b$	$n + n'$

it can be seen that y_a has the hypergeometric distribution $H(y_a + y'_b, n, n')$ under the null hypothesis

$$\mathcal{H}_\lambda : \lambda = 0$$

of no order effects and conditional on the marginal totals. This can be tested using Pearson's chi-square test or Fisher's exact test.

2.4. THE MANTEL–HAENSZEL METHOD

We are often interested only in investigating the relationship between two binary variables (e.g., a disease and an exposure); however, we have to control for confounders. A confounding variable, or a confounder, is a variable that is associated with both the disease and the exposure. For example, in Example 1.1, a case-control study was undertaken to investigate the relationship between lung cancer and employment in shipyards during World War II among male residents of coastal Georgia. In this case, smoking is a potential confounder; it has been found to be associated with lung cancer and it may be associated with employment because construction workers are likely to be smokers. Specifically, we want to know:

(i) among smokers, whether or not Shipbuilding and Lung Cancer are related, and
(ii) among nonsmokers, whether or not Shipbuilding and Lung Cancer are related.

The underlying question is the one concerning conditional independence between Lung Cancer and Shipbuilding; however, we do not want to reach separate conclusions, one at each level of smoking. Assuming that the confounder, Smoking, is not an effect modifier (i.e., smoking does not alter the relationship between lung cancer and shipbuilding), we want to pool data for a combined decision. When both the disease and the exposure are binary, a popular method to achieve this task is the Mantel–Haenszel method (Mantel and Haenszel, 1959). The process can be summarized as follows:

(i) We form 2×2 tables, one at each level of the confounder.
(ii) At a level of the confounder, we have

	Disease Classification		
Exposure	+	−	Total
+	a	b	r_1
−	c	d	r_2
Total	c_1	c_2	n

Under the null hypothesis and fixed marginal totals, cell $(1,1)$ frequency "a" is distributed as hypergeometric with mean and variance,

$$E_0(a) = \frac{r_1 c_1}{n} \qquad \mathrm{Var}_0(a) = \frac{r_1 r_2 c_1 c_2}{n^2(n-1)},$$

and the Mantel–Haenszel test is based on the z-statistic

$$z = \frac{\sum a - \sum \frac{r_1 c_1}{n}}{\sqrt{\sum \frac{r_1 r_2 c_1 c_2}{n^2(n-1)}}}$$

where the summation ($\sum$) is across levels of the confounder.

Since we assume that the counfounder is not an effect modifier, the odds ratio is constant accross its levels. The odds ratio at each level is estimated by ad/bc; the Mantel–Haenszel procedure pools data across levels of the confounder to obtain a combined estimate:

$$\widehat{\mathrm{OR}}_{\mathrm{MH}} = \frac{\sum \frac{ad}{n}}{\sum \frac{bc}{n}}.$$

Example 2.11. A case-control study was conducted to identify reasons for the exceptionally high rate of lung cancer among male residents of coastal Georgia (Blot et al., 1978). The primary risk factor under investigation was employment in shipyards during World War II, and data are tabulated separately for three levels of smoking as follows:

Smoking	Shipbuilding	Cases	Controls
No	Yes	11	35
	No	50	203
Moderate	Yes	70	42
	No	217	220
Heavy	Yes	14	3
	No	96	50

There are three 2×2 tables, one for each level of smoking; in Example 1.1, the last two tables were combined and presented together for simplicity.

We begin with the 2×2 table for nonsmokers:

Shipbuilding	Cases	Controls	Total
Yes	$11(a)$	$35(b)$	$46(r_1)$
No	$50(c)$	$203(d)$	$253(r_2)$
Total	$61(c_1)$	$238(c_2)$	$299(n)$

We have, for the nonsmokers,

$$a = 11$$

$$\frac{r_1 c_1}{n} = \frac{(46)(61)}{299}$$

$$= 9.38$$

$$\frac{r_1 r_2 c_1 c_2}{n^2(n-1)} = \frac{(46)(253)(61)(238)}{(299)^2(298)}$$

$$= 6.34 \qquad \text{and}$$

$$\frac{ad}{n} = \frac{(11)(203)}{299}$$

$$= 7.47,$$

$$\frac{bc}{n} = \frac{(35)(50)}{299}$$

$$= 5.85.$$

The process is repeated for each of the other two smoking levels. For moderate smokers,

$$a = 70$$

$$\frac{r_1 c_1}{n} = \frac{(112)(287)}{549}$$

$$= 58.55$$

$$\frac{r_1 r_2 c_1 c_2}{n^2(n-1)} = \frac{(112)(437)(287)(262)}{(549)^2(548)}$$

$$= 22.28 \qquad \text{and}$$

$$\frac{ad}{n} = \frac{(70)(220)}{549}$$

$$= 28.05,$$

$$\frac{bc}{n} = \frac{(42)(217)}{549}$$
$$= 16.60$$

and for heavy smokers

$$a = 14$$
$$\frac{r_1c_1}{n} = \frac{(17)(110)}{163}$$
$$= 11.47$$
$$\frac{r_1r_2c_1c_2}{n^2(n-1)} = \frac{(17)(146)(110)(53)}{(163)^2(162)}$$
$$= 3.36 \qquad \text{and}$$
$$\frac{ad}{n} = \frac{(14)(50)}{163}$$
$$= 4.29,$$
$$\frac{bc}{n} = \frac{(3)(96)}{163}$$
$$= 1.77.$$

These results are combined to obtain the z-score

$$z = \frac{(11-9.38)+(70-58.55)+(14-11.47)}{\sqrt{6.34+22.28+3.36}}$$
$$= 2.76$$

and a z-score of 2.76 yields a one-tailed p-value of .0029, which is significant beyond the 1 percent level. This result is stronger than those for tests at each level because it is based on more information where all data at all three smoking levels are used. The combined odds ratio estimate is

$$\widehat{\text{OR}}_{\text{MH}} = \frac{7.47 + 28.05 + 4.29}{5.85 + 16.60 + 1.77}$$
$$= 1.64,$$

representing an approximate increase of 64 percent in lung cancer risk for those employed in the shipbuilding industry.

Note: An SAS program would include these instructions:

```
DATA;
INPUT SMOKE SHIP CANCER COUNT;
CARDS;
1 1 1 11
1 1 2 35
1 2 1 50
1 2 2 203
2 1 1 70
2 1 2 42
2 2 1 217
2 2 2 220
3 1 1 14
3 1 2 3
3 2 1 96
3 2 2 50;
PROC FREQ;
WEIGHT COUNT;
TABLES SMOKE*SHIP*CANCER/CMH;
```

The result is given in a chi-square form ($X^2 = 7.601$, $p = .006$); CMH stands for Cochran–Mantel–Haenszel statistic.

Example 2.12. A case-control study was conducted to investigate the relationship between myocardial infarction (MI) and oral contraceptive use (OC) (Shapiro et al., 1979). The data, stratified by cigarette smoking, were

Smoking	OC Use	Cases	Controls
No	Yes	4	52
	No	34	754
Yes	Yes	25	83
	No	171	853

An application of the Mantel–Haenszel procedure yields

	Smoking	
	No	Yes
a	4	25
$\frac{r_1 c_1}{n}$	2.52	18.70
$\frac{r_1 r_2 c_1 c_2}{n^2(n-1)}$	2.25	14.00
$\frac{ad}{n}$	3.57	18.84
$\frac{bc}{n}$	2.09	12.54

The combined z-score is

$$z = \frac{(4 - 2.52) + (25 - 18.70)}{\sqrt{2.25 + 14.00}}$$
$$= 1.93,$$

which is significant at the 5 percent level (one-tailed). The combined odds ratio estimate is

$$\widehat{\text{OR}}_{\text{MH}} = \frac{3.57 + 18.84}{2.09 + 12.54}$$
$$= 1.53,$$

representing an approximate increase of 53 percent in myocardial infarction for oral contraceptive users.

2.5. INFERENCES FOR GENERAL TWO-WAY TABLES

Perhaps the most frequent use of the chi-square distribution is to test the null hypothesis that two categorical variables are independent. Two variables are independent if the distribution of one is the same no matter what the level of the other. For example, if socioeconomic status and area of residence are independent, we would expect to find the same proportions of families in the low, medium, and high socioeconomic groups in all areas of the city.

2.5.1. Comparison of Several Proportions

This is a straight extension of the method for the comparison of two proportions, an analog of the one-way *analysis of variance* (ANOVA) procedure.

In this type of problem, we have k independent samples of binary data $(n_1,x_1),(n_2,x_2),\ldots,(n_k,x_k)$, where the n's are sample sizes and the x's are the numbers of positive outcomes in the k samples. For these k independent binomial samples, we consider the null hypothesis

$$\mathcal{H}_0 : \pi_1 = \pi_2 = \cdots = \pi_k$$

expressing the equality of the k population proportions. Let

$$p_i = x_i/n_i; \qquad i = 1,2,\ldots,k$$

be the sample proportion of group i and p be the "pooled proportion" defined by

$$p = \frac{x_1 + x_2 + \cdots + x_k}{n_1 + n_2 + \cdots + n_k}.$$

The test statistic is given by the following formula,

$$\chi^2 = \frac{\sum n_i(p_i - p)^2}{p(1-p)},$$

where the summation is over the k groups and the decision is made referring to the chi-square distribution with $(k-1)$ degrees of freedom. If we apply this procedure to the comparison of two proportions, then we would have the same result as if we were to use the method of section 2.3.1.

Example 2.13. A study was undertaken to investigate the roles of blood-borne environmental exposures on ovarian cancer from assessment of consumption of coffee, tobacco, and alcohol (Whittemore et al., 1988). Study subjects consist of 188 women in the

San Francisco Bay area with epithelial ovarian cancers diagnosed in 1983–1985, and 539 control women. Of the 539 controls, 280 were hospitalized women without overt cancer, and 259 were chosen from the general population by random telephone dialing. Data for coffee consumption are summarized as follows:

Coffee Drinkers	Cases	Hospital Controls	Population Controls	Total
Yes	177	249	233	659
No	11	31	26	68
Total	188	280	259	727

In this example, we have:

$$k = 3$$

$$n_1 = 189, \qquad x_1 = 177$$

$$p_1 = \frac{177}{189} = .937$$

$$n_2 = 280, \qquad x_2 = 249$$

$$p_2 = \frac{249}{280} = .865$$

$$n_3 = 259, \qquad x_3 = 233$$

$$p_3 = \frac{233}{259} = .900$$

$$p = \frac{659}{727} = .906$$

leading to

$$\chi^2 = \frac{189(.937 - .906)^2 + 280(.865 - .906)^2 + 259(.900 - .906)^2}{(.906)(1 - .906)} = 0.662.$$

The difference between the three groups is not significant; coffee consumption does not seem to be associated with epithelial ovarian cancer.

2.5.2. Testing for Independence in Two-Way Tables

Data in a $2 \times k$ table may come from different sampling models, including the case of independent binomial samples (called product binomial) of the previous section. For example, we may have two multinomial samples (a case-control study with a multinomial risk factor) or one multinomial sample with $2k$ categories (a cross-sectional survey). If data come from a sampling model other than the product binomial, or if the contingency table has more than two rows and more than two columns, then the method of the previous section does not seem to apply because we simply do not have k proportions to compare. In these general cases of an $I \times J$ table, however, we can use the $\{\widehat{m_{ij}}\}$ in tests of independence through the Pearson's chi-square statistic,

$$X^2 = \sum_{i,j} \frac{(x_{ij} - \widehat{m_{ij}})^2}{\widehat{m_{ij}}},$$

or the likelihood ratio chi-square statistic:

$$G^2 = 2 \sum_{i,j} x_{ij} \log \frac{x_{ij}}{\widehat{m_{ij}}}.$$

For large samples, both X^2 and G^2 have approximately a chi-squared distribution with df degrees of freedom under the null hypothesis of independence,

$$df = (I - 1)(J - 1);$$

greater values of the test statistics lead to a rejection of H_0. This is applicable regardless of sampling model; in fact, as applied to the data coming from a product binomial, the Pearson's chi-square X^2 yields the same result as that of the previous section.

Example 2.14. The following table shows the results of a survey; each subject of a sample of 300 adults was asked to indicate which of three policies they favored with respect to smoking in public places (data were taken from Daniel, 1987). The numbers in parentheses are expected frequencies.

	Policy Favored				
Highest Education Level	No Restrictions on Smoking	Smoking Allowed in Designated Areas Only	No Smoking at All	No Opinion	Total
College Graduate	5(8.75)	44(46)	23(13.25)	3(4.5)	75
High School	15(17.5)	100(92)	30(26.5)	5(9)	150
Grade School	15(8.75)	40(46)	10(13.25)	10(4.5)	75
Total	35	184	53	18	300

An application of the Pearson's chi-square test, at six (6) degrees of freedom, yields

$$X^2 = \frac{(5-8.75)^2}{8.75} + \frac{(44-46)^2}{46} + \cdots + \frac{(10-4.5)^2}{4.5}$$

$$= 25.50.$$

The result indicates a high correlation between education levels and preferences about smoking in public places ($p = .001$). Similar results can be obtained using the likelihood ratio chi-square test ($G^2 = 20.60$, $p = .002$).

Note: An SAS program would include these instructions:

```
DATA;
INPUT EDUCAT POLICY COUNT;
CARDS;
1 1 5
1 2 44
1 3 23
1 4 3
2 1 15
```

```
2 2 100
2 3 30
2 4 5
3 1 15
3 2 40
3 3 10
3 4 10;
PROC FREQ;
WEIGHT COUNT;
TABLES EDUCAT*POLICY/CHISQ;
```

2.5.3. Ordered $2 \times k$ Contingency Tables

This section presents an efficient method for use with ordered $2 \times k$ contingency tables, tables with 2 rows and with k columns having a certain natural ordering. Let us first consider an example concerning the use of seat belts in automobiles. Each accident in this example is classified according to whether a seat belt was used and the severity of injuries received: none, minor, major, or death. The data were as follows:

	Extent of Injury Received			
Seat Belt	None	Minor	Major	Death
Yes	75	160	100	15
No	65	175	135	25

To compare the extent of injury in those who used seat belts and those who did not, we can perform a chi-square test as presented in the previous section. Such an application of the chi-square test yields

$$X^2 = 9.26$$

with 3 degrees of freedom ($0.01 \le p \le 0.05$). Therefore, the difference between the two groups is significant at the 5 percent level but not at the 1 percent level. However, the usual chi-square calculation takes no account of the fact that the extent of injury has a natural ordering: none < minor < major < death. In addition, the percent of seat belt users in each injury group decreases from level "none" to level "death":

None: $75/(75+65) = 58\%$

Minor: $160/(100+175) = 48\%$

Major: $100/(100+135) = 43\%$

Death: $15/(15+25) = 38\%$

We now present a special procedure specifically designed to detect such a "trend" and will use the same example to show that it attains a higher degree of significance. In general, consider an ordered $2 \times k$ table with frequencies:

	Column Level				
Row	1	2	⋯	k	Total
1	a_1	a_2	⋯	a_k	A
2	b_1	b_2	⋯	b_k	B
Total	n_1	n_2	⋯	n_k	N

The number of "concordances" is calculated by

$$C = a_1(b_2 + \cdots + b_k) + a_2(b_3 + \cdots + b_k) + \cdots + a_{k-1}b_k.$$

(The term "concordance" pair as used in the above example corresponds to a less severe injury for the seat belt user.) The number of "discordances" is

$$D = b_1(a_2 + \cdots + a_k) + b_2(a_3 + \cdots + a_k) + \cdots + b_{k-1}a_k.$$

In order to perform the test, we calculate the statistic

$$S = C - D,$$

then standardize it to obtain

$$z = \frac{S - \mu_S}{\sigma_D}$$

where $\mu_S = 0$ is the mean of S under the null hypothesis and

$$\sigma_S = \left\{ \frac{AB}{3N(N-1)} [N^3 - n_1^3 - n_2^3 - \cdots - n_k^3] \right\}^{1/2}.$$

The standardized z-score is distributed as standard normal if the null hypothesis is true. The null hypothesis is rejected at the 5 percent level if

$$z < -1.96 \qquad \text{or} \qquad z > 1.96.$$

Example 2.15. For the above study on the use of seat belts in automobiles, we have

$$\begin{aligned} C &= 75(175+135+25)+160(135+25)+(100)(25) \\ &= 53{,}225 \\ D &= 65(160+100+15)+175(100+15)+(135)(15) \\ &= 40{,}025. \end{aligned}$$

In addition, we have

$$\begin{aligned} A &= 350 \\ B &= 390 \\ n_1 &= 130 \\ n_2 &= 335 \\ n_3 &= 235 \\ n_4 &= 40 \\ N &= 740. \end{aligned}$$

Substituting these values into the equations of the test statistic, we have

$$\begin{aligned} S &= 53{,}225 - 40{,}025 \\ &= 13{,}200 \end{aligned}$$

$$\sigma_S = \left\{ \frac{(350)(390)}{(3)(740)(739)} [740^3 - 130^3 - 335^3 - 235^3 - 40^3] \right\}^{1/2}$$
$$= 5414.76$$

leading to

$$z = 13{,}200/5414.76$$
$$= 2.44,$$

which shows a higher degree of significance (one-tailed p-value = 0.0073) than that of the chi-square test.

The method seems ideal for the evaluation of ordinal risk factors in case-control studies. In this case, the statistic

$$\theta = C/D$$

serves as a generalized odds ratio measuring the effect of the exposure.

Example 2.16. Prematurity, which ranks as the major cause of neonatal morbidity and mortality, has traditionally been defined on the basis of a birth weight under 2500g. But this definition encompasses two distinct types of infants: infants who are small because they are born early, and infants who are born at or near term but are small because their growth was retarded. "Prematurity" has now been replaced by

(i) "low birth weight" to describe the second type
(ii) "preterm" to characterize the first type (babies born before 37 weeks of gestation)

A case-control study of the epidemiology of preterm delivery was undertaken at Yale–New Haven Hospital in Connecticut during 1977 (Berkowitz, 1981). The study population consisted of 175 mothers of singleton preterm infants and 303 mothers of singleton full-term infants. The following table gives the distribution of age of the mother:

Age	Cases	Controls	Total
14–17	15	16	31
18–19	22	25	47
20–24	47	62	109
25–29	56	122	178
≥ 30	35	78	113
Total	175	303	

We have:

$$\begin{aligned} C &= 15(25+62+122+78)+22(62+122+78) \\ &\quad +47(122+78)+(56)(78) \\ &= 20{,}911 \\ D &= 16(22+47+56+35)+25(47+56+35) \\ &\quad +(62)(56+35)+(122)(35) \\ &= 15{,}922 \\ S &= 20{,}911-15{,}922 \\ &= 4989 \\ \sigma_S &= \left\{ \frac{(175)(303)}{(3)(478)(477)}[478^3-31^3-47^3-109^3-178^3-113^3] \right\}^{1/2} \\ &= 2794.02 \end{aligned}$$

leading to

$$\begin{aligned} z &= 4989/27.94.02 \\ &= 1.79, \end{aligned}$$

which shows a significant association between the mother's age and preterm delivery (one-tailed p-value = 0.0367); the younger the mother, the more likely the preterm delivery.

Suppose we want to evaluate an ordinal risk factor in the presence of a confounder assuming that the confounder is not an effect modifier. The data can be presented in a series of ordered $2 \times k$ tables, one at each level of the confounder. For example, the data on the mother's age in Example 2.16 may be tabulated separately for the races, one ordered 2×5 table for whites and one for nonwhites. At a level of the confounder, we can apply the above method to obtain the statistic S and its standard error under the null hypothesis, σ_S. The combined decision is then based on

$$z = \frac{\sum\{S - \mu_S\}}{\sum\{\sigma_D\}},$$

where the summation is across levels of the confounder. The statistic z is distributed as standard normal if the null hypothesis is true; so that the null hypothesis is rejected at the two-sided 5 percent level if

$$z < -1.96 \qquad \text{or} \qquad z > 1.96.$$

This can be considered as a generalized Mantel–Haenszel procedure; it reduces to the Mantel–Haenszel test if the risk factor has two levels. The statistic

$$\theta = \frac{\sum(C/N)}{\sum(D/N)}$$

serves as a generalized odds ratio which reduces to $\mathrm{OR}_{\mathrm{MH}}$ if the risk factor has two levels.

2.6. SAMPLE SIZE DETERMINATION

The determination of the size of a sample is a crucial element in the design of a survey or a clinical trial. In designing any study, one of the first questions that must be answered is, "How large must the sample be to accomplish the goals of the study?" Depending on the study goals, the planning of sample size can be approached in two different ways: either in terms of controlling the width of a desired confidence interval for the parameter of interest, or in terms of controlling the risk of making Type II errors.

Suppose the goal of another study is to estimate an unknown population proportion π, say, the smoking rate of a certain well-defined population. For the confidence interval to be useful, it must be short enough to pinpoint the value of the parameter reasonably well with a high degree of confidence. If a study is unplanned or poorly planned, there is a real possibility that the resulting confidence interval will be too long to be of any use to the researcher. In this case, we may determine to have an error of the estimate not exceeding d. The 95 percent confidence interval for the population proportion π is

$$p \pm 1.96\sqrt{\frac{p(1-p)}{n}},$$

where p is the sample proportion. Therefore, our goal is expressed as

$$1.96\sqrt{\frac{p(1-p)}{n}} \leq d,$$

leading to the required minimum sample size

$$n = \frac{(1.96)^2 p(1-p)}{d^2}$$

(rounded up to the next integer). This required sample size is affected by three factors:

(i) the degree of confidence, i.e., 95 percent which yields the coefficient, 1.96;

(ii) the maximum tolerated error, d, determined by the investigator(s); and

(iii) the proportion p itself.

This third factor is unsettling! In order to find n so as to obtain an accurate value of the proportion, we need the proportion itself. There is no perfect, exact solution for this. Usually, we can use information from similar studies, past studies, or studies on similar populations. If no good prior knowledge about the proportion is available, we can replace $p(1-p)$ by .25 and use a "conservative" sample size

$$n_c = \frac{(1.96)^2(.25)}{d^2}$$

because $n_c \geq n$ regardless of the value of π.

Let us consider now the problem where we want to design a study to compare two proportions. For example, a new vaccine will be tested in which subjects are to be randomized into two groups of equal size: a control (unimmunized) group (group 1), and an experimental (immunized) group (group 2). Subjects, in both control and experimental groups, will be challenged by a certain type of bacteria and we wish to compare the infection rates. The null hypothesis to be tested is

$$\mathcal{H}_0 : \pi_1 = \pi_2$$

versus

$$\mathcal{H}_A : \pi_1 < \pi_2.$$

How large a total sample should be used to conduct this vaccine study?

Suppose that it is important to detect a reduction of infection rate

$$d = \pi_2 - \pi_1.$$

If we decide to preset the size of the study at $\alpha = .05$ and want the power $(1-\beta)$ to detect the difference d, then the required sample size is given by this complicated formula:

$$N = \frac{4[2z_{1-\alpha}\pi(1-\pi) + z_{1-\beta}\{\pi_1(1-\pi_1) + \pi_2(1-\pi_2)\}]^2}{(\pi_2 - \pi_1)^2}.$$

In this formula, the quantities $z_{1-\alpha}$ and $z_{1-\beta}$ are defined as in the previous section; π is the common value of the proportions under $\mathcal{H}_0$. It is obvious that the problem of planning sample size is more difficult and a good solution requires a deeper knowledge of the scientific problem: some good idea of the magnitude of the proportions π_1, π_2 themselves.

Example 2.17. A new vaccine will be tested in which subjects are to be randomized into two groups of equal size: a control (unimmunized) group and an experimental (immunized) group. Based on prior knowledge about the vaccine through small pilot studies, the following assumptions are made:

(i) The infection of the control group (when challenged by a certain type of bacteria) is expected to be about 50%, i.e.,

$$\pi_2 = .50.$$

(ii) About 80% of the experimental group is expected to develop adequate antibodies (that is, at least a twofold increase). If antibodies are inadequate, then the infection rate is about the same as for a control subject. But if an experimental subject has adequate antibodies, then the vaccine is expected to be about 85% effective (that corresponds to a 15% infection rate against the challenged bacteria).

Putting these assumptions together, we obtain an expected value of π_1:

$$\begin{aligned}\pi_1 &= (.80)(.15) + (.20)(.50) \\ &= .22.\end{aligned}$$

Suppose also that we decide to preset $\alpha = .05$ and want the power to be about 90% (i.e., $\beta = .10$). In other words, we use

$$\begin{aligned}z_{1-\alpha} &= 1.96 \\ z_{1-\beta} &= 1.65.\end{aligned}$$

From this information, the required total sample size is

$$\begin{aligned}N &= \frac{4[(2)(1.96)(.50)(.50) + (1.65)\{(.50)(.50) + (.22)(.78)\}]^2}{(.50 - .22)^2} \\ &\cong 144\end{aligned}$$

so that each group will have 72 subjects. In this solution we use

$$\pi = .50$$

because under the null hypothesis, the vaccine is not effective (or is only as effective as the control) so that the common value of the infection rate is that of the control group, 50%.

2.7. EXERCISES

1. The relationship between prior condom use and tubal pregnancy was assessed in a population-based case-control study at Group Health Cooperative of Puget Sound during 1981–86 (Li et al., 1990). The results are:

		Cases	Controls
Condom	Never	176	488
Use	Ever	51	186

Compare the proportions of condom users in the two groups, cases vs. controls.

2. Epidemic keratoconjunctivitis (EKC) or "shipyard eye" is an acute infectious disease of the eye. A case of EKC is defined as an illness

- consisting of redness, tearing, and pain in one or both eyes for more than three days' duration,
- diagnosed as EKC by an ophthalmologist.

In late October 1977, one (Physician A) of the two ophthalmologists providing the majority of specialized eye care to the residents of a central Georgia county (population 45,000) saw a 27-year-old nurse who had returned from a vacation in Korea with severe EKC. She received symptomatic therapy and was warned that her eye infection could spread to others; nevertheless, numerous cases of an illness similar to hers soon occurred in the patients and staff of the nursing home (Nursing Home A) where she worked (these individuals came to Physician A for diagnosis and treatment). The following table provides exposure history of 22 persons with EKC between October 27, 1977 and January 13, 1978 (when the outbreak stopped after proper control techniques were initiated) (D'Angelo et al., 1981). Nursing Home B, included in this table, is the only other area chronic-care facility.

Exposure Cohort	Number Exposed	Number of Cases
Nursing Home A	64	16
Nursing Home B	238	6

Compare the disease rates of the two nursing homes.

3. The following table provides data taken from a study on the association between race and use of medical care by adults experiencing chest pain in the past year (Strogatz, 1990):

Response	Black	White
MD Seen in Past Year	35	67
MD Seen, Not in Past Year	45	38
MD Never Seen	78	39
Total	158	144

Test for the independence between response and gender.

4. Consider the following data (Begg and McNeil, 1988):

	Tuberculosis		
Xray	No	Yes	Total
Negative	1,739	8	1,747
Positive	51	22	73
Total	1.790	30	1,820

Test for the independence between Xray result and tuberculosis.

5. An important characteristic of glaucoma, an eye disease, is the presence of classical visual field loss. Tonometry is a common form of glaucoma screening wherein, for example, an eye is classified as positive if it has an intraocular pressure of 21 mm Hg or higher at a single reading. Given the following data (Hollows and Graham, 1966),

		Test Result		
		Positive	Negative	
Field	Yes	13	7	20
Loss	No	413	4,567	4,980

calculate the sensitivity and specificity and their 95 percent confidence intervals.

6. In the course of selecting controls for a study to evaluate effect of caffeine-containing coffee on the risk of myocardial infarction among women 30–49 years of age, a study noted appreciable differences in coffee consumption among hospital patients admitted for illnesses not known to be related to coffee use. Among potential controls, the coffee consumption of patients who had been admitted to hospital by conditions having an acute onset (such as fractures) was compared to that of patients admitted for chronic disorders (Rosenberg et al., 1981):

	Cups of Coffee per Day			
Admission by	0	1–4	≥ 5	Total
Acute Conditions	340	457	183	980
Chronic Conditions	2440	2527	868	5835

Test for the independence between admission reason and coffee consumption.

7. A case-control study of the epidemiology of preterm delivery was undertaken at Yale–New Haven Hospital in Connecticut during 1977 (Berkowitz, 1981). The study population consisted of 175 mothers of singleton preterm infants and 303 mothers of singleton full-term infants. Data on the mother's age were given in Example 2.16; the following table gives the distribution of the mother's socioeconomic status:

Socioeconomic Level	Cases	Controls
Upper	11	40
Upper-Middle	14	45
Middle	33	64
Lower-Middle	59	91
Lower	53	58
Unknown	5	5

Test against the trend that the poorer the mother the more likely the preterm delivery.

8. The role of menstrual and reproductive factors in the epidemiology of breast cancer has been reassessed using pooled data from three large case-control studies of breast cancer from several Italian regions (Negri et al., 1988). The following are summarized data for age at menopause and age at first live birth:

Age at First Live Birth	Cases	Controls
< 22	621	898
22–24	795	909
25–27	791	769
≥ 28	1043	775
Age at Menopause	Cases	Controls
< 45	459	543
45–49	749	803
≥ 50	1378	1167

Test against the trend in age at first live birth and in age at menopause.

9. Data were collected from 2,197 white ovarian cancer patients and 8,893 white controls in 12 different U.S. case-control studies conducted by various investigators in the period 1956–1986 (Whittemore et al., 1992). These were used to evaluate the relationship of invasive epithelial ovarian cancer to reproductive and menstrual characteristics, exogenous estrogen use, and prior pelvic surgeries. The following are parts of the data; apply a proper test for each ordered two-way table (state your alternative hypotheses):

(a)

Duration of Unprotected Intercourse (years)	Cases	Controls
< 2	237	477
2–9	166	354
10–14	47	91
≥ 15	133	174

(b)

History of Infertility	Cases	Controls
No	526	966
Yes		
No Drug Use	76	124
Drug Use	20	11

10. Postmenopausal women who develop endometrial cancer are on the whole heavier than women who do not develop the disease. One possible explanation is that heavy women are more exposed to endogenous estrogens which are produced in postmenopausal women by conversion of steroid precursors to active estrogens in peripheral fat. In the face of varying levels of endogenous estrogen production one might ask whether the carcinogenic potential of exogenous estrogens would be the peripheral fat. In the face of varying levels of endogenous estrogen production one might ask whether the carcinogenic potential of exogenous estrogens would be the same in all women. A study has been conducted to examine the relation between weight, replacement estrogen therapy, and endometrial cancer in a case-control study (Kelsey et al., 1982):

		Estrogen Replacement	
Weight (kg)		Yes	No
< 57	Cases	20	12
	Controls	61	183
57–75	Cases	37	45
	Controls	113	378
> 75	Cases	9	42
	Controls	23	140

Use the Mantel–Haenszel test to investigate the effect of estrogen replacement on endometrial cancer.

CHAPTER 3

Loglinear Models

Topics in Chapter 2 focused mainly on the relationship between two categorical factors, a response and an explanatory variables; however, incorrect conclusions may result from investigating variables two at a time. The Mantel–Haenszel method reaches a little further, allowing us to adjust for a confounder, and it has been a very popular methodology. It is a simple procedure to pool the results from a number of two-by-two tables; however, because of this simplicity, the scopes of the Mantel–Haenszel method are rather narrow. Both main factors are binary, only the confounder can have several levels. It also assumes that the odds ratio (OR) between the two main factors remains constant across levels of the confounder, that is, to assume that there are no effect modifications: The confounder does not modify the effect, say, of the explanatory factor on the response, an assumption that cannot be tested by that method alone (or any other methods in Chapter 2). In addition, the purpose of most research is to assess relationships among a set of several categorical variables, some of which may have more than two levels. To achieve this high level of sophistication, we now turn our attention to *loglinear models*, a popular methodology for modeling multidimensional contingency tables.

Loglinear models describe association patterns among categorical variables without specifying their roles; which one is the response and which ones are explanatory. However, when it is natural to treat one variable as the response and the others as explanatory or independent variables, modifications are simple and the resulting loglinear models are equivalent to the logistic models of Chapter 4. The following are two typical examples that will be used again and again as illustrations throughout this chapter, in addition to Example 1.1.

Example 3.1. The data are from an epidemiologic study following an outbreak of food poisoning that occurred at an outing held for the personnel of an insurance company, taken and slightly modified from Bishop et al. (1975) (also see Korff et al., 1952). Of the food eaten, interest focused on potato salad and crabmeat. The variables are: (1) presence or absence of Illness, (2) Potato Salad (eaten or not eaten), and (3) Crabmeat (eaten or not eaten).

	Food Eaten			
	Crabmeat		No Crabmeat	
Consumer's	Potato Salad		Potato Salad	
Illness	Yes	No	Yes	No
Ill	120	4	22	1
Not Ill	80	31	24	23

Example 3.2. Although cervical cancer is not a major cause of death among American women, it has been suggested that virtually all such deaths are preventable; the basic approach to prevention involves cytologic screening using the Papanicolaou (Pap) test. Yet still many thousands of women die from cervical cancer each year! The data from the 1973 National Health Interview Survey (Kleinman

and Kopstein, 1981) are used to examine the relationship between Pap testing and four socioeconomic variables. The five variables are: (1) Pap testing (tested or not tested), (2) Age (25–44, 45–64, and 65+), (3) Residence (metropolitan or nonmetropolitan), (4) Income (poor or nonpoor), and (5) Race (white or black; data for other ethnic groups are not included).

				Pap Test	
Age	Residence	Income	Race	No	Yes
25–44	Metropolitan	Poor	White	77	516
			Black	57	344
		Nonpoor	White	476	5796
			Black	47	701
	Nonmetropolitan	Poor	White	63	387
			Black	41	94
		Nonpoor	White	211	3186
			Black	24	118
54–64	Metropolitan	Poor	White	163	376
			Black	86	172
		Nonpoor	White	859	5646
			Black	121	400
	Nonmetropolitan	Poor	White	160	328
			Black	59	68
		Nonpoor	White	491	2361
			Black	44	72
65+	Metropolitan	Poor	White	449	499
			Black	70	66
		Nonpoor	White	903	1544
			Black	63	70
	Nonmetropolitan	Poor	White	396	365
			Black	71	43
		Nonpoor	White	459	698
			Black	21	10

3.1. LOGLINEAR MODELS FOR TWO-WAY TABLES

Even though we know how to analyze data in two-way tables using the methods of Chapter 2, the aim of this section is to introduce loglinear models for this simple case of two categorical variables; the models will be then generalized to the case of three dimensions in Section 3.2, which also includes the process of using sample data to fit the models and make inferences. Suppose there is a multinomial sample of size n over the $N = IJ$ cells of an $I \times J$ contingency table; the first factor has I levels represented by rows and the second factor has J levels represented by columns. The cell probabilities $\{\pi_{ij}\}$ for that multinomial distribution form the joint distribution of the two categorical variables. If we define

$$l_{ij} = \log \pi_{ij},$$

then we can write

$$l_{ij} = \lambda + \lambda_{1(i)} + \lambda_{2(j)} + \lambda_{12(ij)}$$

by an analog with the analysis of variance (ANOVA) model. In this formulation,

(i) The first term, λ, is the grand mean of the logs of the probabilities,

$$\lambda = \frac{l_{++}}{IJ}$$

where the plus sign (+) denotes the total when summing across levels of the corresponding factor.

(ii) $\lambda + \lambda_{1(i)}$ is the mean of the logs of the probabilities of the first factor when it is at level i,

$$\frac{\sum_j \log(\pi_{ij})}{J},$$

so that $\lambda_{1(i)}$ the deviation from the grand mean λ:

$$\lambda_{1(i)} = \frac{l_{i+}}{J} - \frac{l_{++}}{IJ}.$$

Thus, it satisfies

$$\sum_i \lambda_{1(i)} = 0$$

and is influenced only by the marginal distribution of the first factor. There are $(I - 1)$ of these terms and they are often of no intrinsic interest, representing only the *main effects* of that factor. Similarly, the main effects of the other factor are represented by $\lambda_{2(j)}$, and there are $(J - 1)$ of these terms.

(iii) The remaining component,

$$\lambda_{12(ij)} = l_{ij} - \frac{l_{i+}}{J} - \frac{l_{+j}}{I} + \frac{l_{++}}{IJ},$$

can be regarded as measures of departures from the independence of two factors. For example, if $I = J = 2$, then it can be shown that

$$\begin{aligned}\lambda_{12(11)} &= \tfrac{1}{4}\log\left(\frac{\pi_{11}\pi_{22}}{\pi_{12}\pi_{21}}\right)\\ &= \tfrac{1}{4}\log(\text{Odds Ratio})\\ &= 0 \quad \text{if} \quad \text{Odds Ratio} = 1.\end{aligned}$$

Since

$$\sum_i \lambda_{12(ij)} = \sum_j \lambda_{12(ij)} = 0,$$

there are $(I - 1)(J - 1)$ of these *interaction terms* the number is called *the degrees of freedom* in testing for the null hypothesis of independence.

In the case of a general two-way table, numerical values of these $\lambda_{12(ij)}$ would indicate where the interaction is strong; the negative or positive sign is not important, reflecting only the arbitrary coding. Of course, the results may be trivial because we can reach the same conclusion by simply inspecting the cell probabilities. However, it is extemely useful after we generalize the models for use with higher-dimensional tables.

3.2. LOGLINEAR MODELS FOR THREE-WAY TABLES

In a typical study, even if we are interested only in the relationship between a response and an explanatory variable, we still have to control for at least one confounder that can influence the relationship under investigation. Therefore, we end up studying at least three factors simultaneously; and when dealing with three factors, the relationships among them are far more complicated than in the case of two factors. For example, we may want to see if:

(i) The three factors are *mutually independent*. That is, whether

$$\Pr(X_1 = i,\ X_2 = j,\ X_3 = k) = \Pr(X_1 = i)\Pr(X_2 = j)\Pr(X_3 = k)$$

or if

(ii) One factor, say X_3, is *jointly independent* of the other two factors. That is, whether

$$\Pr(X_1 = i,\ X_2 = j,\ X_3 = k) = \Pr(X_1 = i,\ X_2 = j)\Pr(X_3 = k)$$

or if

(iii) Two factors, say X_1 and X_2, are *conditionally independent* given the third factor. That is, whether

$$\Pr(X_1 = i, X_2 = j \mid X_3 = k)$$
$$= \Pr(X_1 = i \mid X_3 = k)\Pr(X_2 = j \mid X_3 = k).$$

The last item, the concept of conditional independence, is very important; it is often the major aim of an epidemiological study. For example, in Example 1.1 of Chapter 1, a case-control study was undertaken to identify reasons for the exceptionally high rate of lung cancer among male residents of coastal Georgia. The exposure under investigation, Shipbuilding, refers to employment in shipyards during World War II; Smoking is only a *confounder* that we want to control for. Specifically, we want to know whether or not Shipbuilding and Lung Cancer are related: (i) among smokers, and (ii) among nonsmokers. The underlying question is the question concerning conditional independence. In another example, Example 3.1 of this chapter, we want to know, for example, if Eating Crabmeat and Food Poisoning are related, considering separately the people who ate potato salad and the people who did not eat potato salad. Again, it is a question of conditional independence. It is interesting and important to note that it is *not* the same as the marginal independence, the one we learn when investigating two factors at a time. For example, from the following data for a two-center trial to compare two treatments:

Center	Outcome	Treatment A	Treatment B
X	Good	162	80
	Bad	38	20
Y	Good	11	21
	Bad	89	179

we can see that Treatment and Outcome are conditionally independent (ORs of 1.07 and 1.05) but not marginally independent (OR of 2.68). This example appears to fit *Simpson's paradox* (see Chapter 1), where the marginal strong association between Outcome and Treatment (with Treatment B showing more effective results) is in a different direction from their weak partial association (at each center, Treatment A is slightly more effective).

With the loglinear approach, we model cell probabilities or, equivalently, *cell counts or frequencies* in a contingency table in terms of association among the variables. Suppose there is a multinomial sample of size n over the $N = IJK$ cells of an $I \times J \times K$ contingency table (I, J, and K are the number of categories for the three factors involved). Then the loglinear models for two-way tables of previous section can be generalized and expressed for three-way tables as follows,

$$l_{ijk} = \lambda + \lambda_{1(i)} + \lambda_{2(j)} + \lambda_{3(k)}\lambda_{12(ij)} + \lambda_{13(ik)} + \lambda_{23(jk)} + \lambda_{123(ijk)},$$

subjected to similar constraints (i.e., summing across indices to zero). The (full or saturated) model decomposes the log of cell probability π_{ijk} (or equivalently, the log of expected cell frequency, m_{ijk}, because the expected cell frequency can be expressed as $m_{ijk} = n\pi_{ijk}$) into:

(i) A constant λ.

(ii) Terms representing main effects, $\lambda_{1(i)}$, $\lambda_{2(j)}$, and $\lambda_{3(k)}$. There are $(I - 1)$, $(J - 1)$, and $(K - 1)$ of these main effects terms for the three factors; they are only influenced by the marginal distributions of the three factors and are, therefore, often of no intrinsic interests.

(iii) Terms representing two-factor interactions, $\lambda_{12(ij)}$, $\lambda_{13(ik)}$, and $\lambda_{23(jk)}$; there are $(I-1)(J-1)$, $(I-1)(J-1)$, and $(J-1) \times (K-1)$ of these.

(iv) Terms used as measures of three-factor interaction, $\lambda_{123(ijk)}$. There are $(I-1)(J-1)(K-1)$ of these three-factor interaction terms.

If the terms in the last group are not zero, i.e., three-factor interaction is present, the presence or absence of a factor would modify the relationship between the other two factors. For example, in the context of Example 1.1 of Chapter 1, if the three-factor interaction is present, the odds ratio relating Lung Cancer to Shipbuilding calculated from nonsmokers would be different from the odds ratio relating Lung Cancer to Shipbuilding calculated from smokers. Similarly, the relationship between Lung Cancer and Smoking varies from one level of Shipbuilding to another. The Mantel–Haenszel method of Chapter 2 assumes that these three-factor interaction terms are absent.

3.2.1. The Models of Independence

Starting with the saturated model, we can translate a null hypothesis (or a concept of independence) into a loglinear model by setting certain of the above λ-terms equal to zero. The various concepts of independence for three-way tables can be grouped as follows:

Type of Independence	Symbol	Loglinear Model
Mutual Independence	(X_1, X_2, X_3)	$\lambda + \lambda_1 + \lambda_2 + \lambda_3$
Joint Independence	(X_1, X_2X_3)	$\lambda + \lambda_1 + \lambda_2 + \lambda_3 + \lambda_{23}$
from 2 factors	(X_2, X_1X_3)	$\lambda + \lambda_1 + \lambda_2 + \lambda_3 + \lambda_{13}$
	(X_3, X_1X_2)	$\lambda + \lambda_1 + \lambda_2 + \lambda_3 + \lambda_{12}$
Conditional	(X_1X_3, X_2X_3)	$\lambda + \lambda_1 + \lambda_2 + \lambda_3 + \lambda_{13} + \lambda_{23}$
Independence	(X_1X_2, X_2X_3)	$\lambda + \lambda_1 + \lambda_2 + \lambda_3 + \lambda_{12} + \lambda_{23}$
	(X_1X_2, X_1X_3)	$\lambda + \lambda_1 + \lambda_2 + \lambda_3 + \lambda_{12} + \lambda_{13}$
No Three-Factor Interaction	(X_1X_2, X_1X_3, X_2X_3)	$\lambda + \lambda_1 + \lambda_2 + \lambda_3 + \lambda_{12} + \lambda_{13} + \lambda_{23}$

(Subscripted factor levels are dropped for simplicity, e.g., λ_1 is used for $\lambda_{1(i)}$.) For example, referring to the lung cancer data of Example 1.1, the null hypothesis that Lung Cancer (L) and Shipbuilding (B) are conditionally independent given Smoking (S) is expressed as:

$$H_0 : \lambda + \lambda_B + \lambda_L + \lambda_S + \lambda_{BS} + \lambda_{LS}.$$

As for the food poisoning data of Example 3.1, the null hypothesis that both Crabmeat (C) and Potato Salad (P) have nothing to do with the Illness (I) is expressed as:

$$H_0 : \lambda + \lambda_C + \lambda_I + \lambda_P + \lambda_{CP}.$$

3.2.2. Relationships between Terms and Hierarchy of Models

Loglinear model term λ_A is a lower-order relative of term λ_B if A is a subset of B. For example, λ_2 is a lower-order relative of λ_{23} and λ_{12} is a lower-order relative of λ_{123}. If λ_A is a lower-order relative of λ_B, then λ_B is a higher-order relative of λ_A. For example, λ_2 is a lower-order relative of λ_{23}, and λ_{123} is a higher-order relative of λ_{23}. A loglinear model is a hierarchical model if:

(i) a λ-term is zero, then all of its higher-order relatives are zero, and

(ii) a λ-term is not zero, then all of its lowerer-order relatives are not zero.

For example, all eight models of independence for three-way tables of the previous section are hierarchical. However, we have not considered all possible variants of our loglinear models. For example, we have not considered the model

$$H : \lambda + \lambda_1 + \lambda_2 + \lambda_3 + \lambda_{123},$$

partly because it does not have any meaningful interpretation, partly because it is a *nonhierarchical model*. Statistical methodology (for example, the estimation of expected frequencies) and corresponding computer software is currently available only for hierarchical models. Therefore, we will consider only hierarchical models for higher-dimensional tables.

3.2.3. Testing a Specific Model

Given the data in a three-way table, we have two different types of statistical inferences:

- **(i)** To test for a specific model; for example, we may want to know whether two specified factors are conditionally independent given the third factor.
- **(ii)** To search for a model that can best explain the relationship(s) found in the observed data.

This section deals with the first issue, where the loglinear model under investigation is the result of a process starting with a research question which leads to a relationship between factors, and finally a null hypothesis.

Expected Frequencies

Expected frequencies are cell counts obtained under the null hypothesis. Given the null hypothesis, these expected frequencies (only for hierarchical models) can be easily estimated. For example, if we consider the model of conditional independence between X_1 and X_2, (X_1X_3, X_2X_3), or

$$\Pr(X_1 = i,\ X_2 = j \mid X_3 = k) = \Pr(X_1 = i \mid X_3 = k)\Pr(X_2 = j \mid X_3 = k),$$

then it can be shown that

$$\widehat{m_{ijk}} = \frac{x_{i+k}x_{+jk}}{x_{++k}}$$

where the x's are observed cell counts and the plus sign indicates a summation accross the index. However, in practice these tedious jobs should be left to computer packaged programs, such as SAS; we will show some samples of computer programs in subsequent examples.

Test Statistic

When measuring goodness-of-fit in two-way tables, we often rely on the Pearson's chi-quare statistic:

$$\sum_{i,j,k} \frac{(x_{ijk} - \widehat{m_{ijk}})^2}{\widehat{m_{ijk}}}.$$

However, for a technical reason, the so-called *partition of chi-squares* as seen later, we compare the observed frequencies and the expected frequencies in higher-dimensional tables using the likelihood ratio chi-square statistic:

$$\chi^2 = 2 \sum_{i,j,k} x_{ijk} \log \frac{x_{ijk}}{\widehat{m_{ijk}}}.$$

Degree of Freedom

The degree of freedom for the above likelihood ratio chi-square statistic is equal to the number of λ-terms which are set equal to zero in the model being tested. For example, if we want to test the model of conditional independence between X_1 and X_2,

$$H_0 : \lambda + \lambda_1 + \lambda_2 + \lambda_3 + \lambda_{13} + \lambda_{23},$$

for which we set λ_{12} and λ_{123} equal to zero, the degree of freedom is

$$df = (I - 1)(J - 1) + (I - 1)(J - 1)(K - 1).$$

Similarly, if we want to test for the model of no three-factor interaction,

$$H_0 : \lambda + \lambda_1 + \lambda_2 + \lambda_3 + \lambda_{12} + \lambda_{13} + \lambda_{23},$$

the degree of freedom is

$$df = (I - 1)(J - 1)(K - 1).$$

(This last result is needed, for example, in testing the assumption of the Mantel–Haenszel method.)

Example 3.3. Refer to the lung cancer data of Example 1.1 and suppose we are interested in the null hypothesis that Lung Cancer (L) and Shipbuilding (B) are conditionally independent:

$$H_0 : \lambda + \lambda_B + \lambda_L + \lambda_S + \lambda_{BS} + \lambda_{LS}.$$

Fitting the above model, we have the result,

$$\chi^2 = 6.09 \quad \text{with} \quad 2 \text{ dfs}; \qquad p = .0477,$$

indicating that those two factors are related; workers are more likely to have lung cancer.

Note: An SAS program would include these instructions:

```
INPUT CANCER SHIP SMOKING COUNT;
CARDS;
1 1 1 203
1 1 2 270
1 2 1 35
1 2 2 45
2 1 1 50
2 1 2 313
2 2 1 11
2 2 2 84:
PROC CATMOD:
WEIGHT COUNT;
MODEL CANCER*SHIP*SMOKING = RESPONSE/NODESIGN;
LOGLIN CANCER SHIP SMOKING CANCER*SMOKING SHIP*
  SMOKING;
```

Example 3.4. Refer to the food poisoning data of Example 3.1 and suppose we are interested in the null hypothesis that both Crabmeat (C) and Potato Salad (P) have nothing to do with the Illness (I):

$$H_0 : \lambda + \lambda_C + \lambda_I + \lambda_P + \lambda_{CP}.$$

Fitting the above model, we have a highly significant result,

$$\chi^2 = 62.05 \quad \text{with} \quad 3 \text{ dfs}; \qquad p < .0001,$$

indicating that at least one of the two items are related to the food poisoning outbreak.

3.2.4. Measures of Association

Consider the case of three categorical variables X_1, X_2, and X_3 with data presented in the form of a three-way table. Let X_1 be the response variable with two levels (1 if event of interest is observed, e.g., being a case; 2 if event of interest is not observed, e.g., being a control), X_2 be the explanatory factor (maybe with more than two categories), and X_3 be the confounder. Suppose we are interested in comparing the effect of X_2 (on the response X_1) when X_2 is at level i with the effect of the same factor when X_2 is at level l. For example, in the context of Example 1.1, we want to compare the odds of having lung cancer (X_1) for those employed in the shipbuilding industry (level i of X_2) with the odds of having lung cancer (X_1) for those not employed in the shipbuilding industry (level l of X_2). Of course, we must control for smoking (X_3). Assuming that X_3 is at level k, the above odds ratio (OR) is given by the formula:

$$\log \mathrm{OR}_{i \text{ vs. } l} = 2(\lambda_{12(1i)} - \lambda_{12(1l)}) + 2(\lambda_{123(1ik)} - \lambda_{123(1lk)}).$$

In the special case of no effect modification, the above formula is simplified to

$$\log \mathrm{OR}_{i \text{ vs. } l} = 2(\lambda_{12(1i)} - \lambda_{12(1l)}),$$

and if the explanatory factor has only two levels (1 if exposed and 2 if not exposed), then we have the odds ratio associated with the exposure

$$\log \mathrm{OR} = 4\lambda_{12(11)}$$

because of the constraint for the model

$$\lambda_{12(11)} + \lambda_{12(12)} = 0.$$

Example 3.5. Refer to the food poisoning data of Example 3.1 and suppose we are interested in the effect of Potato Salad (eaten vs. not eaten) on the Illness and are considering testing the model:

$$H_0 : \lambda + \lambda_{\text{C}} + \lambda_{\text{I}} + \lambda_{\text{P}} + \lambda_{\text{CP}} + \lambda_{\text{IP}}.$$

This is the model of conditional independence between Crabmeat and Illness, and the result,

$$\chi^2 = 3.30 \quad \text{with} \quad 2 \text{ dfs;} \qquad p = .192,$$

indicates a good fit. In addition, we obtained from fitting this acceptable model,

$$\lambda_{\text{IP}(11)} = 0.6727,$$

leading to an odds ratio of:

$$\begin{aligned} \text{OR} &= \exp[(4)(0.6727) \\ &= 14.74. \end{aligned}$$

This result means that the odds of having the illness for those who ate potato salad is about 14.74 times the odds of having the illness for those who did not eat potato salad. It can seen that the above result is quite similar to that obtained using the Mantel–Haenszel method. An application of the Mantel–Haenszel procedure would proceed as follows.

(i) From the data for people who ate crabmeat:

	Potato Salad		
Illness	Yes	No	Total
Yes	120	4	124
No	80	31	111
Total	200	35	235

we have

$$\frac{ad}{n} = \frac{(120)(31)}{235}$$

$$\frac{bc}{n} = \frac{(80)(4)}{235}.$$

(ii) From the data for people who did not eat crabmeat:

	Potato Salad		
Illness	Yes	No	Total
Yes	22	1	23
No	24	23	47
Total	46	24	70

we have

$$\frac{ad}{n} = \frac{(22)(23)}{70}$$

$$\frac{bc}{n} = \frac{(24)(1)}{70}.$$

Therefore:

$$\mathrm{OR}_{\mathrm{MH}} = \frac{\dfrac{(120)(31)}{235} + \dfrac{(22)(23)}{70}}{\dfrac{(80)(4)}{235} + \dfrac{(24)(1)}{70}}$$

$$= 13.53.$$

Example 3.6. The following table provides data relating infant mortality to the duration of the gestation and the smoking status of the mother (Wermuth, 1976):

	Gestation (G; days)			
	260 or less		More than 260	
	Smoking (S)		Smoking (S)	
Infant Death (D)	Yes	No	Yes	No
Yes	13	91	7	38
No	11	462	583	5606

Suppose we are interested in the effect of premature delivery on infant mortality after controlling for the mother's smoking status. The model of conditional independence to be tested is

$$H_0 : \lambda + \lambda_G + \lambda_S + \lambda_D + \lambda_{GS} + \lambda_{DS}.$$

Fitting the above model, we have

$$\chi^2 = 2.26 \quad \text{with} \quad 2 \text{ dfs}; \qquad p = 0.3231,$$

indicating a good fit. In addition, we obtained from fitting this model

$$\lambda_{DG(11)} = 0.8320,$$

leading to an odds ratio of

$$\begin{aligned} \text{OR} &= \exp[(4)(0.8320)] \\ &= 27.88. \end{aligned}$$

This result means that the odds of having an infant death among premature deliveries is about 27.88 times the odds of having an infant death among full-term deliveries. We should note that the result depends on the fitted model; it is conventional to use the smallest good-fitted model which contains the interaction term under investigation. For example, the conditional independence model of Example 3.6 contains an insignificant term, the interaction between gestation and smoking ($p = 0.4640$). Hence, we prefer to estimate the above odds ratio measuring the effect of premature delivery on infant mortality by fitting the smaller model,

$$H_0 : \lambda + \lambda_G + \lambda_S + \lambda_D + \lambda_{DS},$$

even though the results may be very close.

3.2.5. Searching for the Best Model

As mentioned earlier, given the data in a three-way table, we have two different types of statistical inferences:

(i) To test for a specific model; for example, we may want to know whether two specified factors are conditionally independent given the third factor.
(ii) To search for the best model that can explain the relationship(s) found in the observed data.

A loglinear model H_2 is *nested* in a model H_1 if every nonzero λ-term in H_2 is also contained in H_1 ($H_2 < H_1$). For example, if we denote

$$\begin{aligned} H_1 &= \lambda + \lambda_1 + \lambda_2 + \lambda_3 + \lambda_{12} + \lambda_{13} + \lambda_{23} \\ H_2 &= \lambda + \lambda_1 + \lambda_2 + \lambda_3 + \lambda_{13} + \lambda_{23} \\ H_3 &= \lambda + \lambda_1 + \lambda_2 + \lambda_3 + \lambda_{23}, \end{aligned}$$

then

(i) $H_3 < H_2$, and
(ii) $H_2 < H_1$.

In this nested hierarchy of models, it can be shown that if

$$H_3 < H_2 < H_1,$$

then the likelihood ratio chi-square statistics satisfy the reversed inequality, i.e.:

$$\chi^2(H_1) < \chi^2(H_2) < \chi^2(H_3).$$

(One of the reasons we do not use the Pearson goodness-of-fit is that this property does not necessarily hold for every set of nested models). Furthermore, it can be shown that if a model H_2 is *nested* in a model H_1, then

$$\chi^2 = \chi^2(H_2) - \chi^2(H_1)$$

is distributed as chi-square with

$$df = df(H_2) - df(H_1).$$

(This also does not necessarily hold for Pearson's chi-squares). This result has at least two important applications, the second of which provides the necessary framework for selecting the best model.

Application 1

If

(i) a model H_1 fits (i.e., $p \gg .05$, say), and
(ii) $H_1 - H_2$ is *one* term, say λ_θ,

we can use $\chi^2(H_2) - \chi^2(H_1)$ as a test statistic for

$$H_0 : \lambda_\theta = 0.$$

Example 3.7. Refer to the food poisoning data of Example 3.1 and suppose we are interested *only* in the effect of Potato Salad (eaten vs. not eaten) on the Illness, i.e., testing against

$$H_0 : \lambda_{IP} = 0.$$

We found:

(i) The model of no three-term interaction

$$H_1 = \lambda + \lambda_I + \lambda_C + \lambda_P + \lambda_{IC} + \lambda_{IP} + \lambda_{CP}$$

fits ($\chi^2(H_1) = .27$, 1 df).

(ii) Let H_2 be the model of conditional independence,

$$H_2 = \lambda + \lambda_{\mathrm{I}} + \lambda_{\mathrm{C}} + \lambda_{\mathrm{P}} + \lambda_{\mathrm{IC}} + \lambda_{\mathrm{CP}} +,$$

which is only one term smaller than H_1, the term under investigation (λ_{IP}).

By fitting H_2 we obtained $\chi^2(H_2) = 47.62$, 2 dfs. The test for

$$H_0 : \lambda_{\mathrm{IP}} = 0$$

corresponds to $\chi^2 = 47.35$ with 1 df, which is highly significant. It can be seen that the above result is quite similar to that obtained using the Mantel–Haenszel method. An application of the Mantel–Haenszel procedure would proceed as follows.

(i) From the data for people who ate crabmeat:

	Potato Salad		
Illness	Yes	No	Total
Yes	120	4	124
No	80	31	111
Total	200	35	235

We have for the top-left cell:

$$\text{Observed value} = 120$$

$$\text{The mean} = \frac{(200)(124)}{235}$$

$$= 105.5$$

$$\text{The variance} = \frac{(200)(35)(124)(111)}{235^2(235-1)}$$

$$= 7.46.$$

(ii) From the data for people who did not eat crabmeat:

	Potato Salad		
Illness	Yes	No	Total
Yes	22	1	23
No	24	23	47
Total	46	24	70

We have for the top left-cell:

$$\text{Observed value} = 22$$

$$\text{The mean} = \frac{(46)(23)}{70} = 15.1$$

$$\text{The variance} = \frac{(46)(24)(23)(47)}{70^2(70-1)} = 3.53.$$

Therefore:

$$\chi^2_{MH} = \frac{\{(120 - 105.5) + (22 - 15.1)\}^2}{7.46 + 3.53} = 41.67.$$

The difference is: The Mantel–Haenszel method assumes that H_1 (the model of no three-term interaction) fits, whereas we *know* that H_1 fits (after testing it).

Application 2

If

(i) model H_2 is nested in model H_1 and $H_1 - H_2$ is *one* term, say λ_θ, and

(ii) both H_1 and H_2 fit, and

(iii) the test for

$$H_0 : \lambda_\theta = 0$$

is not significant,

then H_2 is *better than* or *preferred over* H_1. The *best* model is the most preferred model among the hierarchy of good-fitted models. In other words, the best model fits and contains only significant terms. It should be noted that, unfortunately, there is more than one best model because there may be more than one hierarchy of good-fitted models.

Example 3.8. Refer to the Food poisoning data of Example 3.1; by fitting all eight models of independence to this three-way table we obtain:

Model	Description	χ^2	df
H_4	$\lambda + \lambda_I + \lambda_C + \lambda_P$	68.16	4
H_{3a}	$\lambda + \lambda_I + \lambda_C + \lambda_P + \lambda_{IC}$	59.42	3
H_{3b}	$\lambda + \lambda_I + \lambda_C + \lambda_P + \lambda_{IP}$	15.10	3
H_{3c}	$\lambda + \lambda_I + \lambda_C + \lambda_P + \lambda_{CP}$	56.36	3
H_{2a}	$\lambda + \lambda_I + \lambda_C + \lambda_P + \lambda_{IP} + \lambda_{IC}$	6.36	2
H_{2b}	$\lambda + \lambda_I + \lambda_C + \lambda_P + \lambda_{IC} + \lambda_{CP}$	47.62	2
H_{2c}	$\lambda + \lambda_I + \lambda_C + \lambda_P + \lambda_{IP} + \lambda_{CP}$	3.30	2
H_1	$\lambda + \lambda_I + \lambda_C + \lambda_P + \lambda_{IC} + \lambda_{IP} + \lambda_{CP}$	.27	1

There is one hierarchy consisting of two nested good models:

$$H_{2c} < H_1.$$

Since the single term separating them, λ_{IC}, is not significant ($\chi^2 =$ 3.03, 1 df), model H_{2c} is *better* and, therefore, the best model to explain the food poisoning data set.

3.2.6. Collapsing Tables

Consider the case of three categorical variables X_1, X_2, and X_3 with data presented in the form of a three-way table and suppose that we are interested only in the relationship between X_1 and X_2. If we collapse the cross-classification over X_3, yielding a two-way table, we would like to know whether, for example, the odds ratio obtained from this marginal table is correct (i.e., the same that would result from the loglinear model method of subsection 3.2.4). By an analogy with the method for correlation coefficients for continuous data, it can be shown that (Bishop, et al., 1975):

> In a three-dimensional table the interaction between two variables may be measured from the marginal table by collapsing over the third variable if and only if the third variable is independent of at least one of the two variables exhibiting the interaction.

Example 3.9. Refer to the food poisoning data of Example 3.1. Since the best model we found in Example 3.8,

$$H_{2c} : \lambda + \lambda_{\text{I}} + \lambda_{\text{C}} + \lambda_{\text{P}} + \lambda_{\text{IP}} + \lambda_{\text{CP}},$$

indicates that only Crabmeat and Illness are independent,

(i) we can collapse on Crabmeat in studying the relationship between Potato Salad and Illness, but

(ii) we cannot collapse on Potato Salad in order to study the relationship between Crabmeat and Illness.

However, this seems an unnecessary question: In order to answer it, we need to perform a thorough loglinear model analysis; and after performing the needed loglinear model analysis, we no longer need to collapse the table!

3.3. LOGLINEAR MODELS FOR HIGHER-DIMENSIONAL TABLES

Although all the models and methods discussed in section 3.2 have been in the context of three-way tables, their extensions to higher-dimensional tables (four-way tables, five-way tables, etc.) are relatively straightforward. Given the data in a cross-classified table, regardless of the dimension, we still have two different types of statistical inferences:

(i) To test for a specific model; for example, we may want to know whether two specified factors are conditionally independent given all the other factors.

(ii) To search for a model that can best explain the relationship(s) found in the observed data.

3.3.1. Testing a Specific Model

The process is still the same as that in subsection 3.2.3 for three-way tables: estimation of expected frequencies under the null hypothesis, calculation of the goodness-of-fit statistic (likelihood ratio chi-square), and determination of the degree of freedom. Of course, we implement these steps using the same computer packaged program, SAS PROC CATMOD. The key and only needed step is to describe the loglinear model to be tested starting with a research question which leads to a relationship between factors, and finally a null hypothesis. We are also limited to considering only hierarchical models.

Suppose we consider the cervical screening data of Example 3.2 with all five variables: Age (A), Income (I), Race (R), Metropolitan (M; yes/no), and Pap testing (P). The null hypothesis of no Income effects, for example, can be investigated in two different ways:

(i) Unconditional approach: Here we test the hierarchical model H_0 not containing the basic term under investigation, λ_{IP}.

(ii) Conditional approach: A two-step procedure. In step 1, we fit the same model H_0 as in the unconditional approach; in step 2, we fit another model, H_1, by adding in the term λ_{IP}, then use the process for nested models focusing on $\chi^2(H_0) - \chi^2(H_1)$ with $df(H_0) - df(H_1)$. This conditional approach is like an extension of the Mantel–Haenszel method. In order to use this approach, H_1 must fit because we assume that there are no other factors that would modify the effects of Income on Pap testing (terms such as λ_{AIP}; otherwise, look for a good-fitted model containg λ_{IP}).

Example 3.10. Refer to the cervical screening data of Example 3.2. By fitting the model

$$H_0 : \lambda + \lambda_A + \lambda_I + \lambda_M + \lambda_R + \lambda_{AR} + \lambda_{MR} + \lambda_{IR} + \lambda_{RP} + \lambda_{AM}$$
$$+ \lambda_{AI} + \lambda_{AP} + \lambda_{MI} + \lambda_{MP} + \lambda_{AMR} + \lambda_{AIR} + \lambda_{ARP}$$
$$+ \lambda_{MIR} + \lambda_{MRP} + \lambda_{AMI} + \lambda_{AMP} + \lambda_{AMIR} + \lambda_{AMRP},$$

we obtain: $\chi^2 = 317.28$ with 12 degrees of freedom ($p < .0001$). Model H_1 does not fit, which rules out the conditional approach.

3.3.2. Searching for the Best Model

As mentioned earlier, given the data in a contingency table, we have two different types of statistical inferences:

(i) To test for a specific model.
(ii) To search for a model that can best explain the relationship(s) found in the observed data.

The test for a specific model is conducted in the same way regardless of the dimension of the table. However, the search for the best model is different for higher-dimensional tables. In the case of three-way tables, we fit all eight possible models of independence; there is usually one hierarchy of nested models from which to pick the best model. For higher-dimensional tables, it is impossible to do that; for a four-way table there are 113 possible hierarchical models, and for a ten-way table the number of hierarchical models is 3,475, 978! Therefore, we need a more efficient strategy.

The most commonly used procedure is a stepwise procedure. The description here is for four-way tables, for simplicity, but can be extended to tables with five or more dimensions. This is, in fact, a two-step process:

Step 1: In the first step, we look for a *starting model*, one that is close to the best model. We begin by choosing a significant level, say .05, and then we test for the goodness-of-fit of all models of uniform orders. A model of uniform orders contains all interaction terms involving the same number of factors. For example, there are three models of uniform orders for a four-way table:

$$H_1 : \lambda + \lambda_1 + \lambda_2 + \lambda_3 + \lambda_3$$

$$H_2 : \lambda + \lambda_1 + \lambda_2 + \lambda_3 + \lambda_4 + \lambda_{12} + \lambda_{13} + \lambda_{14} + \lambda_{23} + \lambda_{24} + \lambda_{34}$$

$$H_3 : \lambda + \lambda_1 + \lambda_2 + \lambda_3 + \lambda_4 + \lambda_{12} + \lambda_{13} + \lambda_{14} + \lambda_{23} + \lambda_{24} + \lambda_{34} + \lambda_{123} + \lambda_{124} + \lambda_{134} + \lambda_{234}.$$

If model H_3 does not fit the data, we stop; the best model is the saturated model. If model H_3 fits but model H_2 does not, the best model is somewhere between them, that is, to contain some but not all three-factor interactions; we can choose as the starting model either H_2 or H_3. If both H_2 and H_3 fit but model H_1 does not, the best model is somewhere between H_1 and H_2 and we can use either one as the starting model.

Step 2: In the second step, we apply the stepwise procedure to the starting model found in step 1 in order to reach the best model. For example, if model H_3 fits but model H_2 does not, then:

(i) we can choose H_2 as a starting model and do a forward addition by adding in one three-factor interaction term at a time, or

(ii) we can choose H_3 as a starting model and do a backward elimination by deleting one three-factor interaction term at a time. The process continues until no term can be added or deleted.

Since SAS PROC CATMOD does not have an automatic stepwise option, it is much easier to perform a backward elimination. When we fit the larger model H_3, the computer output also includes significance levels for each three-factor interaction term; we can usually eliminate all nonsignificant terms and reach the best model in one step. We illustrate this process first with a four-way table, then with a five-way table, the cervical screening data of Example 3.2.

Example 3.11. The following data are from a study of the survival of breast cancer patients (taken from Bishop et al., 1975). The main factor is three-year survival status (yes/no). Other factors are: Age (under 50, 50–69, 70 or over), Treatment Center (Tokyo, Boston, and Glamorgan), and two histologic criteria: nuclear grade (malignant appearance/benign appearance) and degree of chronic inflammatory reaction (minimal/moderate-severe). The last two factors were interrelated and together could be regarded as a description of the disease state. Let's call this Inflammation with four categories: (a) minimal with malignant appearance, (b) minimal with benign appearance, (c) moderate-severe with malignant appearance, and (d) moderate-severe with benign appearance. The four factors are: Age (A), Center (C), Inflammation (I), and Survival status (S); the aim is to search for the best model showing the relationship contained in the observed data.

			Inflammation			
Center	Age	Survival	(a)	(b)	(c)	(d)
Tokyo	Under 50	No	9	7	4	3
		Yes	26	68	25	9
	50–69	No	9	9	11	2
		Yes	20	46	18	5
	70 or Over	No	2	3	1	0
		Yes	1	6	5	1
Boston	Under 50	No	6	7	6	0
		Yes	11	24	4	0
	50–69	No	8	20	3	2
		Yes	18	58	10	3
	70 or Over	No	9	18	3	0
		Yes	15	26	1	1
Glamorgan	Under 50	No	16	7	3	0
		Yes	16	20	8	1
	50–69	No	16	12	3	0
		Yes	27	39	10	4
	70 or Over	No	3	7	3	0
		Yes	12	11	4	1

Fitting the above three models of uniform orders:

H_1: all main effects,
H_2: all two-factor interaction terms, and
H_3: all three-factor interaction terms,

we found that H_2 fits ($p = 0.2392$; in fact, there is no need to fit H_3) in this case) but model H_1 does not ($p < .0001$). Therefore, we can use H_2 as a starting model and proceed with a backward elimination.

From fitting model H_2, we obtain:

Term	df	χ^2	p-Value
C*A	4	58.92	$< .0001$
C*S	2	8.51	0.0142
C*I	6	31.49	$< .0001$
A*S	2	4.17	0.1245
A*I	6	1.55	0.9560
S*I	3	1012	0.0176

The results indicate that among the six three-factor interaction terms, two of them are not significant at the 5 percent level: A*S ($p = 0.1245$) and A*I ($p = 0.9560$). By deleting these terms, we obtain the best model:

$$H_{best} = \lambda + \lambda_A + \lambda_C + \lambda_I + \lambda_S + \lambda_{AC} + \lambda_{CS} + \lambda_{CI} + \lambda_{IS}$$

($\chi^2 = 44.12$, 41 dfs, $p = 0.3411$). As far as the survival of patients is concerned, there are two significanr factors: Degree of Inflammation and Treatment Center. Recall the condition for collapsing tables in section 3.2.6; it can be seen that we cannot collapse to two-way tables in order to study the effects of each factor (Age, Inflammation, and Center) on breast cancer survival:

(i) If we are interested in the effects of Age, we cannot collapse on C because of significant terms A*C and C*S.

(ii) If we are interested in the effects of Inflammation, we cannot collapse on C because of significant terms C∗I and C∗S.

(iii) If we are interested in the differences between centers, we cannot col!apse on I because of significant terms C∗I and I∗S.

Example 3.12. Refer to the cervical screening data of Example 3.2 with all five variables: Age (A), Income (I), Race (R), Metropolitan (M; yes/no), and Pap testing (P). Fitting the four models of uniform orders:

H_1: all main effects,
H_2: all two-factor interaction terms,
H_3: all three-factor interaction terms, and
H_4: all four-factor interaction terms,

we found that H_3 fits ($p = 0.1171$; in fact, there is no need to fit H_4) in this case) but model H_2 does not ($p < .0001$). Therefore, we can use H_3 as a starting model and proceed with a backward elimination.

From fitting model H_3, we obtain:

Term	df	χ^2	p-Value
A∗M∗R	2	2.36	0.3077
A∗I∗R	2	10.87	0.0044
A∗R∗P	2	3.36	0.1861
M∗I∗R	1	4.42	0.0356
M∗R∗P	1	24.82	$< .0001$
I∗R∗P	1	10.19	0.0014
A∗M∗I	2	5.96	0.0508
A∗M∗P	2	3.13	0.2096
A∗I∗P	2	37.61	$< .0001$
M∗I∗P	10.27	0.6015	

The results indicate that among the ten three-factor interaction terms, 5 of them are not significant at the 5 percent level: A∗M∗R

($p = 0.3077$), A*R*P ($p = 0.1861$), A*M*I ($p = 0.0508$), A*M*P ($p = 0.2096$), and M*I*P ($p = 0.6015$). By deleting these terms, we obtain the best model:

$$H_0 : \lambda + \lambda_A + \lambda_I + \lambda_M + \lambda_R + \lambda_{AR} + \lambda_{MR} + \lambda_{IR} + \lambda_{RP} + \lambda_{AM} + \lambda_{AI}$$
$$+ \lambda_{AP} + \lambda_{MI} + \lambda_{MP} + \lambda_{AIR} + \lambda_{MIR} + \lambda_{MRP} + \lambda_{IRP} + \lambda_{AIP}.$$

In other words, all four factors (Age, Income, Metropolitan or Residence, and Race) are related to Pap testing. There are also three significant three-factor interaction terms:

(i) Age and Income modify the effects of each other.
(ii) Income and Race modify the efffects of each other.
(iii) Metropolitan and Race modify the effects of each other.

For example, the degree of difference between poor women and nonpoor women, in terms of testing or not testing for cervical cancer, is different when calculating among whites from that when calculating among blacks. This would be much easier to illustrate using the odds ratios as formulated in the next section.

3.3.3. Measures of Association

Consider the case of three categorical variables X_1, X_2, and X_3 with a significant three-factor interaction term $X_1 * X_2 * X_3$. Let X_1 be the response variable with two levels (1 if event of interest is observed, e.g., not being tested; 2 if event of interest is not observed, e.g., being tested in the context of the Cervical Cancer Screening problem), X_2 be the explanatory factor (maybe with more than two categories), and X_3 be the confounder. Suppose we are interested in comparing the effect of X_2 (on the response X_1) when X_2 is at level i with the effect of the same factor when X_2 is at level l. For example, in the context of Example 3.2, we want to compare the odds of not testing for cervical cancer (P) for those aged 70 or over (level 3 of A) to the odd of not testing for cervical cancer (P) for those aged 50 or under (level 1 of A). Of course, we must control for Income (I) because Age and Income modify the effects of each other. Assuming that X_3 is at level k, the above odds ratio is given by the formula:

$$\log \text{OR}_{i \text{ vs. } I} = 2(\lambda_{12(1i)} - \lambda_{12(1I)}) + 2(\lambda_{123(1ik)} - \lambda_{123(1Ik)}).$$

The second term is induced by the effect modification; therefore, there are two such terms if there are two effect modifiers. In the special case of no effect modifications, the above formula is simplified to

$$\log \text{OR}_{i \text{ vs. } I} = 2(\lambda_{12(1i)} - \lambda_{12(1I)}),$$

and if the explanatory factor has only two levels (1 if exposed and 2 if not exposed), then we have the odds ratio associated with the exposure

$$\log \text{OR} = 4\lambda_{12(11)}$$

because of the constraint for the model:

$$\lambda_{12(11)} + \lambda_{12(12)} = 0.$$

Example 3.13. Refer to the breast cancer survival data of Example 3.11, where we found the best model

$$H_{\text{best}} = \lambda + \lambda_{\text{A}} + \lambda_{\text{C}} + \lambda_{\text{I}} + \lambda_{\text{S}} + \lambda_{\text{AC}} + \lambda_{\text{CS}} + \lambda_{\text{CI}} + \lambda_{\text{IS}}.$$

Suppose that we are interested in the difference between two treatment centers, Boston versus Tokyo. By fitting the best model, we obtain the following estimates for λ_{CS}:

	Treatment Centers		
Survival	Tokyo	Boston	Glamorgan
Yes	0.2086	−0.1368	−0.0718
No	−0.2086	0.1368	0.0718

$$\begin{aligned}\log \text{OR}_{\text{Boston vs. Tokyo}} &= 2\{(-0.1368) - (0.2086)\}\\ &= -0.6908\end{aligned}$$

leading to:

$$OR_{\text{Boston vs. Tokyo}} = \exp[(-0.6908)]$$
$$= 0.50.$$

In other words, the treatments in Boston are only half as effective as those in Tokyo.

Example 3.14. Refer to the cervical cancer screening data of Example 3.2 where we found the best model:

$$H_0 : \lambda + \lambda_A + \lambda_I + \lambda_M + \lambda_R + \lambda_{AR} + \lambda_{MR} + \lambda_{IR} + \lambda_{RP} + \lambda_{AM} + \lambda_{AI}$$
$$+ \lambda_{AP} + \lambda_{MI} + \lambda_{MP} + \lambda_{AIR} + \lambda_{MIR} + \lambda_{MRP} + \lambda_{IRP} + \lambda_{AIP}$$

(Example 3.12). Suppose we want to compare the older women (65 or over) versus the youger women (25–44). Of course, we have to control for Income because λ_{AIP} is significant. By fitting the best model, we obtain the following estimates for λ_{AP}:

	Age Group		
Pap Test	25–44	45–64	65 or Over
No	−0.5095	0.0019	0.5076
Yes	0.5095	−0.0019	−0.5076

We also obtain the following estimates for λ_{AIP}:

		Age Group		
Income	Pap Test	25–44	45–64	65 or Over
Poor	No	0.0392	0.0441	−0.0833
	Yes	−0.0392	−0.0441	0.0833
Nonpoor	No	−0.0392	−0.0441	0.0833
	Yes	0.0392	0.0441	−0.0833

For poor women:

$$\log \mathrm{OR}_{\text{65-or-over vs. 25–44}}$$
$$= 2\{(0.5076) - (-.5095)\} + 2\{(-0.0833) - (0.0392)\}$$
$$= 1.7892,$$

leading to:

$$\mathrm{OR}_{\text{65-or-over vs. 25–44}} = \exp(1.7892)$$
$$= 5.985.$$

For nonpoor women:

$$\log \mathrm{OR}_{\text{65-or-over vs. 25–44}}$$
$$= 2\{(0.5076) - (-.5095)\} + 2\{(0.0833) - (-0.0392)\}$$
$$= 2.28,$$

leading to:

$$\mathrm{OR}_{\text{65-or-over vs. 25–44}} = \exp(2.2792)$$
$$= 9.77.$$

In other words, the women in age group 65-or-over are much more likely not tested for cervical cancer as compared to women in age group 25–44:

(i) The odds ratio is 5.85 if they are poor.
(ii) The odds ratio is 9.77 if they are nonpoor.

Example 3.15. Refer to the cervical cancer screening data of Examples 3.2 and 3.14, but this time we want to compare black women versus white women. The calculation is more complicated because there are two effect modifiers; terms λ_{MRP} and λ_{IRP} are both significant at the 5 percent level. By fitting the best model, we obtain the following estimates for λ_{RP}:

	Race	
Pap Test	White	Black
No	−0.1437	0.1437
Yes	0.1437	−0.1437

We also obtain the following estimates for λ_{MRP}:

		Race	
Metropolitan	Pap Test	White	Black
Metro	No	0.0698	−0.0698
	Yes	−0.0698	0.0698
Non-metro	No	−0.0698	0.0698
	Yes	0.0698	−0.0698

and the following estimates for λ_{IRP}:

		Race	
Income	Pap Test	White	Black
Poor	No	0.0444	−0.0444
	Yes	−0.0444	0.0444
Nonpoor	No	−0.0444	0.0444
	Yes	0.0444	−0.0444

For poor-metro women:

$$\begin{aligned}\log \mathrm{OR}_{\text{Blacks vs. Whites}} &= 2\{(0.1437) - (-0.1437)\} + 2\{(-0.0698) - (0.0698)\} \\ &\quad + 2\{(-0.0444) - (0.0444)\} \\ &= 0.118,\end{aligned}$$

leading to:

$$\begin{aligned} OR_{\text{Blacks vs. Whites}} &= \exp(0.118) \\ &= 1.13. \end{aligned}$$

For nonpoor-metro women:

$$\begin{aligned} &\log OR_{\text{Blacks vs. Whites}} \\ &\quad = 2\{(0.1437) - (-0.1437)\} + 2\{(-0.0698) - (0.0698)\} \\ &\qquad + 2\{(0.0444) - (-0.0444)\} \\ &\quad = 0.4732, \end{aligned}$$

leading to:

$$\begin{aligned} OR_{\text{Blacks vs. Whites}} &= \exp(0.118) \\ &= 1.61. \end{aligned}$$

For poor-nonmetro women:

$$\begin{aligned} &\log OR_{\text{Blacks vs. Whites}} \\ &\quad = 2\{(0.1437) - (-0.1437)\} + 2\{(0.0698) - (-0.0698)\} \\ &\qquad + 2\{(-0.0444) - (0.0444)\} \\ &\quad = 0.6764, \end{aligned}$$

leading to:

$$\begin{aligned} OR_{\text{Blacks vs. Whites}} &= \exp(0.118) \\ &= 1.97. \end{aligned}$$

For nonpoor-nonmetro women:

$$\begin{aligned}\log \mathrm{OR}_{\text{Blacks vs. Whites}} &= 2\{(0.1437)-(-0.1437)\}+2\{(0.0698)-(-0.0698)\} \\ &\quad +2\{(0.0444)-(-0.0444)\} \\ &= 1.0316,\end{aligned}$$

leading to:

$$\begin{aligned}\mathrm{OR}_{\text{Blacks vs. Whites}} &= \exp(0.118) \\ &= 2.81.\end{aligned}$$

In other words, black women are more likely not tested for cervical cancer as compared to white women:

(i) The odds ratio is 1.13 if they are poor and from the metropolitan area.
(ii) The odds ratio is 1.61 if they are nonpoor and from the metropolitan area.
(iii) The odds ratio is 1.97 if they are poor and not from the metropolitan area.
(iv) The odds ratio is 2.81 if they are nonpoor and not from the metropolitan area.

3.3.4. Searching for a Model with a Dependent Variable

Loglinear models describe association patterns among categorical variables without specifying their roles; which one is the response and which ones are explanatory factors. However, when it is natural to treat one variable as the response and the others as explanatory or independent variables, modifications have to be made in order to make the resulting loglinear models equivalent to logistic models of Chapter 4. This can be achieved by simply keeping in the models to be fitted all terms relating independent variables with each other; the changes in the numerical results are negligible most of the time.

Example 3.16. Refer to the breast cancer survival data of Example 3.11 with four factors: Age (A), Center (C), Inflammation (I), and Survival status (S); and suppose we are searching for the best model with Survival Status (S) being identified as the response. Without a response variable, the three models of uniform orders are:

H_1: all main effects,
H_2: all two-factor interaction terms, and
H_3: all three-factor interaction terms.

In the presence of the dependent variable (S), these are expanded to include the following terms,

$$\lambda_{\text{AC}} + \lambda_{\text{AI}} + \lambda_{\text{CI}} + \lambda_{\text{ACI}},$$

wherever appropriate. The resulting best model turns out to be

$$H_{\text{best}} = \lambda + \lambda_{\text{A}} + \lambda_{\text{C}} + \lambda_{\text{I}} + \lambda_{\text{S}} + \lambda_{\text{AC}} + \lambda_{\text{CS}} + \lambda_{\text{CI}} + \lambda_{\text{IS}} + \lambda_{\text{AI}} + \lambda_{\text{ACI}},$$

which appears different from the result in Example 3.11. However, most numerical results of interest are still very similar; for example,

$$\begin{aligned} \text{OR}_{\text{Boston vs. Tokyo}} &= \exp(-0.6834) \\ &= 0.51 \end{aligned}$$

as compared to an odds ratio of .50 as in Example 3.11.

3.4. EXERCISES

1. Adult male residents of 13 counties of western Washington state in whom testicular cancer had been diagnosed during 1977–1983 were interviewed over the telephone regarding their history of genital tract conditions, including vasectomy (Strader et al., 1988). For comparison, the same interview was given to a sample of men selected from the population of these counties by dialing telephone numbers at random. The following data are tabulated by religious background.

Religion	Vasectomy	Cases	Controls
Protestant	Yes	24	56
	No	205	239
Catholic	Yes	10	6
	No	32	90
Others	Yes	18	39
	No	56	96

(a) Test for the conditional independence between testicular cancer and vasectomy.

(b) Search for the model that can best explain the relationships found in the observed data; choose to do with or without identifying a dependent variable (i.e., testicular cancer in this exercise).

(c) From the best model found in (b), calculate the odds ratio(s) measuring the strength of the relationship between testicular cancer and vasectomy.

2. It has been hypothesized that dietary fiber decreases the risk of colon cancer, while meats and fats are thought to increase this risk. A large study was undertaken to confirm these hypotheses (Graham et al., 1988). Fiber and fat consumptions are classified as "low" or "high" and data are tabulated separately for males and females as follows ("low" means below median):

	Males		Females	
Diet	Cases	Controls	Cases	Controls
Low fat, high fiber	27	38	23	39
Low fat, low fiber	64	78	82	81
High fat, high fiber	78	61	83	76
High fat, low fiber	36	28	35	27

(a) Test for the conditional independence between colon cancer and diet.

(b) Search for the model that can best explain the relationships found in the observed data; choose to do with or without identifying a dependent variable (i.e., colon cancer in this exercise).

(c) From the best model found in (b), calculate the odds ratio(s) measuring the strength of the relationship between colon cancer and diet.

3. Post-neonatal mortality due to respiratory illnesses is known to be inversely related to maternal age, but the role of young motherhood as a risk factor for respiratory morbidity in infants has not been thoroughly explored. A study was conducted in Tucson, Arizona aimed at the incidence of lower respiratory tract illnesses during the first year of life. In this study, over 1,200 infants were enrolled at birth between 1980 and 1984 and the following data are concerned with wheezing lower respiratory tract illnesses (wheezing LRI) No/Yes (Martinez et al., 1992):

Maternal Age (years)	Boys		Girls	
	No	Yes	No	Yes
< 21	19	8	20	7
21–25	98	40	128	36
26–30	160	45	148	42
> 30	110	20	116	25

(a) Test for the conditional independence between wheezing LRI and maternal age.

(b) Search for the model that can best explain the relationships found in the observed data; choose to do with or without identifying a dependent variable (i.e., wheezing LRI in this exercise).

(c) From the best model found in (b), calculate the odds ratio(s) measuring the strength of the relationship between wheezing LRI and maternal age.

4. Since incidence rates of most cancers rise with age, this must always be considered a confounder. The following are stratified data for an unmatched case-control study (Tuyns et al., 1977). The disease was esophageal cancer among men and the risk factor was alcohol consumption.

Age		Daily Alcohol Consumption 80+ g	0–79 g
25–44	Cases	5	5
	Controls	35	270
45–64	Cases	67	55
	Controls	56	277
65+	Cases	24	44
	Controls	18	129

(a) Test for the conditional independence between esophageal cancer and alcohol consumption.

(b) Search for the model that can best explain the relationships found in the observed data; choose to do with or without identifying a dependent variable (i.e., esophageal cancer in this exercise).

(c) From the best model found in (b), calculate the odds ratio(s) measuring the strength of the relationship between esophageal cancer and alcohol consumption.

5. Postmenopausal women who develop endometrial cancer are on the whole heavier than women who do not develop the disease. One possible explanation is that heavy women are more exposed to endogenous estrogens which are produced in postmenopausal women by conversion of steroid precursors to active estrogens in peripheral fat. In the face of varying levels of endogenous estrogen production one might ask whether the carcinogenic potential of exogenous estrogens would be the peripheral fat. In the face of varying levels of endogenous estrogen production one might ask whether the carcinogenic potential of exogenous estrogens would be the same in all women. A study has been conducted to examine the relation between weight, replacement estrogen therapy, and endometrial cancer in a case-control study (Kelsey et al., 1982):

		Estrogen Replacement	
Weight (kg)		Yes	No
< 57	Cases	20	12
	Controls	61	183
57–75	Cases	37	45
	Controls	113	378
> 75	Cases	9	42
	Controls	23	140

(a) Test for the conditional independence between endometrial cancer and estrogen replacement.

(b) Search for the model that can best explain the relationships found in the observed data; choose to do with or without identifying a dependent variable (i.e., endometrial cancer in this exercise).

(c) From the best model found in (b), calculate the odds ratio(s) measuring the strength of the relationship between endometrial cancer and estogen replacement.

6. Data taken from a study to investigate the effects of smoking on cervical cancer are stratified by the number of sexual partners (Nischan et al., 1988). Results are as follows:

Number of Partners	Smoking	Cancer Yes	Cancer No
Zero or one	Yes	12	21
	No	25	118
Two or more	Yes	96	142
	No	92	150

(a) Test for the conditional independence between cervical cancer and number of partners.

(b) Search for the model that can best explain the relationships found in the observed data; choose to do with or without identifying a dependent variable (i.e., cervical cancer in this exercise).

(c) From the best model found in (b), calculate the odds ratio(s) measuring the strength of the relationship between cervical cancer and number of partners.

7. Cases of poliomyelitis were classified by age, by paralytic status, and by whether the subject had been injected with the Salk vaccine (Chin et al., 1961; data were taken from Agresti, 1990):

Age	Salk Vaccine	Paralysis No	Paralysis Yes
0–4	Yes	20	14
	No	10	24
5–9	Yes	15	12
	No	3	15
10–14	Yes	3	2
	No	3	2
15–19	Yes	7	4
	No	1	6
20–29	Yes	12	3
	No	7	5
40+	Yes	1	0
	No	3	2

(a) Test for the conditional independence between paralysis and Salk vaccine.

(b) Search for the model that can best explain the relationships found in the observed data; choose to do with or without identifying a dependent variable (i.e., paralysis in this exercise).

(c) From the best model found in (b), calculate the odds ratio(s) measuring the strength of the relationship between paralysis and Salk vaccine.

8. The following are data on smoking from a survey of seventh graders (Age: 1 = 12 or younger, 2 = 13 or older; Murray et al., 1987):

				Smoking	
Family Structure	Race	Sex	Age	None	Some
Both Parents	Black	Male	1	27	2
			2	12	2
		Female	1	23	4
			2	7	1
	White	Male	1	394	32
			2	142	19
		Female	1	421	38
			2	94	11
Mother Only	Black	Male	1	18	1
			2	13	1
		Female	1	24	0
			2	4	3
	White	Male	1	48	6
			2	25	4
		Female	1	55	15
			2	13	4

(a) Test for the conditional independence between smoking and family structure.

(b) Test to see if there is a difference in smoking rates between boys and girls.

(c) Test to see if race or sex alter the effect of family structure on smoking.

(d) Search for the model that can best explain the relationships found in the observed data; choose to do with or without identifying a dependent variable (i.e., smoking in this exercise).

(e) From the best model found in (d), calculate the odds ratio(s) measuring the strength of the relationship between smoking and family structure.

(f) From the best model found in (d), calculate the odds ratio(s) measuring the strength of the relationship between smoking and sex.

9. The following are data on drinking from the same survey of seventh graders as of the previous data set (Age: 1 = 12 or younger, 2 = 13 or older; Murray et al., 1987):

				Drinking	
				Smoking	
Family Structure	Mother's Occupation	Sex	Age	None	Some
Both Parents	Other	Male	1	405	44
			2	152	23
		Female	1	454	26
			2	102	8
	White-Collar	Male	1	392	39
			2	130	17
		Female	1	434	39
			2	91	11
Mother Only	Other	Male	1	61	11
			2	37	4
		Female	1	83	8
			2	21	3
	White-Collar	Male	1	104	17
			2	40	9
		Female	1	140	20
			2	32	6

(a) Test for the conditional independence between drinking and family structure.

(b) Test to see if there is a difference in drinking rates between boys and girls.

(c) Test to see if race or sex alter the effect of family structure on drinking.

(d) Search for the model that can best explain the relationships found in the observed data; choose to do with or without identifying a dependent variable (i.e., drinking in this exercise).

(e) From the best model found in (d), calculate the odds ratio(s) measuring the strength of the relationship between drinking and family structure.

(f) From the best model found in (d), calculate the odds ratio(s) measuring the strength of the relationship between drinking and sex.

CHAPTER 4

Logistic Regression Models

The purpose of most research is to assess relationships among a set of variables and regression techniques concern the statistical analysis of such relationships. Research designs may be classified as experimental or observational. Regression analyses are applicable to both types; yet the confidence one has in the results of a study can vary with the research type. In most cases, one variable is usually taken to be the response or dependent variable, that is, a variable to be predicted from or explained by other variables. The other variables are called predictors, or explanatory or independent variables. Choosing an appropriate model and analytical technique depends on the type of the dependent variable under investigation. In a variety of applications, the dependent variable of interest may have only two possible outcomes, and therefore can be represented by an indicator variable taking on values 0 and 1. Consider, for example, an analysis of whether or not business firms have a daycare facility, according to the number of female employees. The dependent variable in this study was defined to have two possible outcomes: (i) the firm has a daycare facility, and (ii) the firm does not have a daycare facility, which may be coded as 1 and 0 respectively. As another example, consider a study of drug use among middle school children, as a function of gender and age of child, family structure (e.g., who is the head of household), and family income. In this study, the dependent variable Y was defined to have two possible outcomes: (i) child uses drugs, and (ii) child does not use drugs. Again, these two outcomes may be coded 1 and 0 respectively.

The above examples, and others, show a wide range of applications in which the dependent variable is dichotomous, and hence may be represented by a variable taking the value 1 with probability π and the value 0 with probability $1 - \pi$. Such a variable is a binomial variable and the model often used to express the probability π as a function of potential independent variables under investigation is the logistic regression model. The logistic model has been used extensively and successfully in the health sciences to describe the probability (or risk) of developing a condition—say, a disease—over a specified time period as a function of certain risk factors $X_1, X_2, \ldots, X_k$. The following is a typical example:

Example 4.1. When a patient is diagnosed as having cancer of the prostate, an important question in deciding on treatment strategy for the patient is whether the cancer has spread to the neighboring lymphnodes. The question is so critical in prognosis and treatment that it is customary to operate on the patient (i.e., perform a laparotomy) for the sole purpose of examining the nodes and removing tissue samples to examine under the microscope for evidence of cancer. However, certain variables that can be measured without surgery are predictive of the nodal involvement; and the purpose of the study presented in Brown (1980) was to examine the data for 53 prostate cancer patients receiving surgery, to determine which of five preoperative variables are predictive of nodal involvement. In particular, the principal investigator was interested in the predictive value of the level of acid phosphatase in blood serum. Table 4.1 presents the complete data set. For each of the 53 patients, there are two continuous independent variables, age at diagnosis and level of serum acid phosphatase ($\times 100$; called "acid"), and three binary variables, Xray reading, pathology reading (grade) of a biopsy of the tumor obtained by needle before surgery, and a rough measure of the size and location of the tumor (stage) obtained by palpation with the fingers via the rectum. For these three binary independent variables a value of one signifies a positive or more serious state and a zero denotes a negative or less serious finding. In addition, the sixth column presents the finding at surgery—the primary binary response or dependent variable Y, a value of one denoting nodal involvement, and a value of zero denoting no nodal involvement found at surgery.

A careful reading of the data reveals, for example, that a positive Xray or an elevated acid phosphatase level, in general, seems likely to be associated with nodal involvement found at surgery. However, predictive values of other variables are not clear and to answer the question, for example, concerning the usefulness of acid phosphatase as a prognostic variable, we need a more detailed analysis before a conclusion can be made.

4.1. SIMPLE REGRESSION ANALYSIS

In this section we will discuss the basic ideas of simple regression analysis when only one predictor or independent variable is available for predicting the response of interest. In the interpretation of the primary parameter of the model, we will discuss both scales of measurement, discrete and continuous, even though in practical applications, the independent variable under investigation is often on a continuous scale.

4.1.1. The Logistic Regression Model

The usual "regression" analysis goal is to describe the "mean" of a dependent variable Y as a function of a set of predictor variables. The logistic regression, however, deals with the case where the basic random variable Y of interest is a dichotomous variable taking the value 1 with probability π and the value 0 with probability $(1 - \pi)$. Such a random variable is called a *point-binomial* or *Ber-*

TABLE 4.1. Prostate Cancer Data

Xray	Grade	Stage	Age	Acid	Nodes
0	1	1	64	40	0
0	0	1	63	40	0
1	0	0	65	46	0
0	1	0	67	47	0
0	0	0	66	48	0
0	1	1	65	48	0
0	0	0	60	49	0
0	0	0	51	49	0
0	0	0	66	50	0
0	0	0	58	50	0
0	1	0	56	50	0
0	0	1	61	50	0
0	1	1	64	50	0
0	0	0	56	52	0
0	0	0	67	52	0
1	0	0	49	55	0
0	1	1	52	55	0
0	0	0	68	56	0
0	1	1	66	59	0
1	0	0	60	62	0
0	0	0	61	62	0
1	1	1	59	63	0
0	0	0	51	65	0
0	1	1	53	66	0
0	0	0	58	71	0
0	0	0	63	75	0
0	0	1	53	76	0
0	0	0	60	78	0
0	0	0	52	83	0
0	0	1	67	95	0
0	0	0	56	98	0
0	0	1	61	102	0
0	0	0	64	187	0
1	0	1	58	48	1
0	0	1	65	49	1
1	1	1	57	51	1
0	1	0	50	56	1
1	1	0	67	67	1
0	0	1	67	67	1
0	1	1	57	67	1

TABLE 4.1. *(Continued)*

Xray	Grade	Stage	Age	Acid	Nodes
0	1	1	45	70	1
0	0	1	46	70	1
1	0	1	51	72	1
1	1	1	60	76	1
1	1	1	56	78	1
1	1	1	50	81	1
0	0	0	56	82	1
0	0	1	63	82	1
1	1	1	65	84	1
1	0	1	64	89	1
0	1	0	59	99	1
1	1	1	68	126	1
1	0	0	61	136	1

nouilli variable and it has the simple discrete probability distribution:

$$\Pr(Y = y) = \pi^y (1 - \pi)^{1-y}; \qquad y = 0, 1.$$

Suppose that for the ith individual of a sample ($i = 1, 2, \ldots, n$), Y_i is a Bernouilli variable with

$$\Pr(Y_i = y_i) = \pi_i^{y_i} (1 - \pi_i)^{1-y_i}; \qquad y_i = 0, 1.$$

The logistic regression analysis assumes that the relationship between π_i and the covariate value x_i of the same individual is described by the logistic function

$$\pi_i = \frac{1}{1 + \exp[-(\beta_0 + \beta_1 x_i)]}, \qquad i = 1, 2, \ldots, n.$$

The basic logistic function is given by

$$f(z) = \frac{1}{1 + e^{-z}},$$

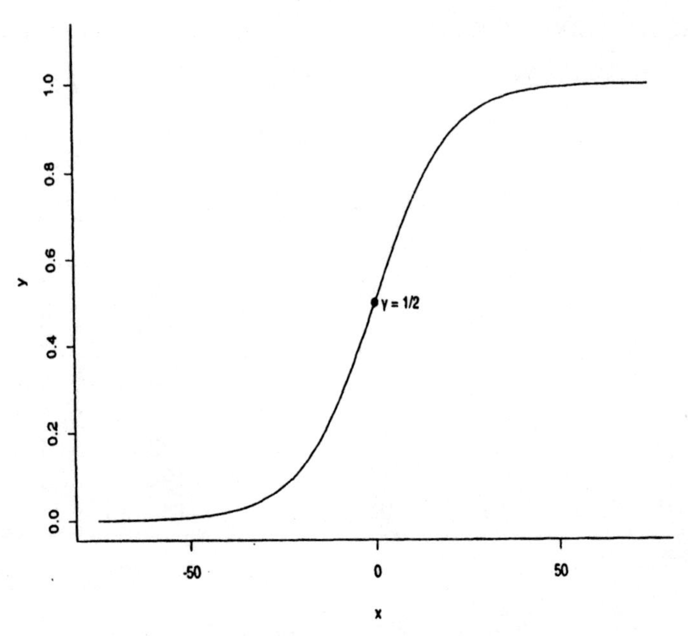

FIGURE 4.1. Basic Logistic Curve

where, as in this simple regression model,

$$z_i = \beta_0 + \beta_1 x_i,$$

or, in the multiple regression model of subsequent sections,

$$z_i = \beta_0 + \sum_{j=1}^{k} \beta_j x_{ji},$$

representing an index of combined risk factors. There are two important reasons that make logistic regression popular:

1. The range of the logistic function is between 0 and 1; that makes it suitable for use as a probability model, representing individual risk.
2. The logistic curve has an increasing S-shape with a threshold; that makes it suitable for use as a biological model, representing risk due to exposure (Figure 4.1).

Under the above simple logistic regression model, the likelihood function is given by

$$L = \prod_{i=1}^{n} \Pr(Y_i = y_i)$$

$$= \prod_{i=1}^{n} \frac{\{\exp[\beta_0 + \beta_1 x_i]\}^{y_i}}{1 + \exp[\beta_0 + \beta_1 x_i]}; \qquad y_i = 0, 1,$$

from which we can obtain "maximum likelihood estimates" of the parameters β_0 and β_1. As previously mentioned, the logistic model has been used both extensively and successfully to describe the probability of developing ($Y = 1$) some disease over a specified time period as a function of a risk factor X.

4.1.2. Measure of Association

Regression analysis serves two major purposes: (1) control or intervention, and (2) prediction. In many studies, such as the one in Example 4.1, one important objective is "measuring" the strength of a statistical relationship between the binary dependent variable and each independent variable or covariate measured from patients; findings may lead to important decisions in patient management (or public health interventions in other examples). In epidemiological studies, such effects are usually measured by the *relative risk* or *odds ratio*; when the logistic model is used, the measure is odds ratio.

For the case of the logistic regression model, the logistic function for the probability π_i can also be expressed as a linear model in the log scale:

$$\ln \frac{\pi_i}{1 - \pi_i} = \beta_0 + \beta_1 x_i.$$

We first consider the case of a binary covariate with the conventional coding

$$X_i = \begin{cases} 0 & \text{if the patient is not exposed} \\ 1 & \text{if the patient is exposed} \end{cases}.$$

Here, the term "exposed" may refer to a risk factor such as smoking, or a patient's characteristic such as race (white/nonwhite) or sex (male/female). It can be seen that, from the above loglinear form of the logistic regression model,

$$\ln(\text{Odd; nonexposed}) = \beta_0$$

$$\ln(\text{Odd; exposed}) = \beta_0 + \beta_1,$$

so that the difference leads, after exponentiating, to

$$e^{\beta_1} = \frac{(\text{Odd; exposed})}{(\text{Odd; nonexposed})}.$$

This represents the odds ratio (OR) associated with the exposure, exposed versus nonexposed. In other words, the primary regression coefficient β_1 is the value of the odds ratio on the log scale.

Similarly, we have for a continuous covariate X and any value x of X,

$$\ln(\text{Odd; } X = x) = \beta_0 + \beta_1(x)$$

$$\ln(\text{Odd; } X = x + 1) = \beta_0 + \beta_1(x + 1),$$

so that the difference leads, after exponentiating, to

$$e^{\beta_1} = \frac{(\text{Odd; } X = x + 1)}{(\text{Odd; } X = x)}.$$

This represents the odds ratio (OR) associated with *one unit increase* in the value of X, $X = x + 1$ versus $X = x$, for example, a systolic blood pressure of 114 mmHg versus 113 mmHg. For m units increase in the value of X, say $X = x + m$ versus $X = x$, the corresponding odds ratio is $e^{m\beta_1}$.

The primary regression coefficient β_1 (and β_0, which is often not needed) can be estimated iteratively using a computer packaged program such as SAS. From the results, we can obtained a point estimate

$$\widehat{\text{OR}} = e^{\hat{\beta}_1}$$

and its 95 percent confidence interval

$$\exp[\hat{\beta}_1 \pm 1.96\text{SE}(\hat{\beta}_1)].$$

4.1.3. The Effect of Measurement Scale

It should be noted that the odds ratio, used as a measure of association between the binary dependent variable and a covariate, depends on the coding scheme for a binary covariate and, for a continuous covariate X, the scale with which to measure X. For example, if we use the following coding for a factor

$$X_i = \begin{cases} -1 & \text{if the subject is not exposed} \\ 1 & \text{if the subject is exposed} \end{cases},$$

then

$$\ln(\text{Odd; nonexposed}) = \beta_0 - \beta_1$$

$$\ln(\text{Odd; exposed}) = \beta_0 + \beta_1,$$

so that we have

$$\begin{aligned} \text{OR} &= \exp[\ln(\text{Odd; exposed}) - \ln(\text{Odd; nonexposed})] \\ &= e^{2\beta_1} \end{aligned}$$

and its 95 percent confidence interval

$$\exp[2(\hat{\beta}_1 \pm 1.96\text{SE}(\hat{\beta}_1))].$$

Of course, the estimate of β_1 under the new coding scheme is only half of that under the former scheme; therefore, the estimate of the OR remains unchanged. The following example, however, will show the clear effect of measurement scale in the case of a continuous measurement.

Example 4.2. Refer to the data for patients diagnosed as having cancer of the prostate in Example 4.1 and suppose we want to investigate the relationship between nodal involvement found at surgery and the level of acid phosphatase in blood serum in two different ways using either (i) $X = \text{Acid}$ or (ii) $X = \log_{10}(\text{Acid})$.

1. For $X = \text{Acid}$, we find

$$\hat{\beta}_1 = .0204,$$

from which the odds ratio for (Acid = 100) versus (Acid = 50) would be

$$\begin{aligned} \text{OR} &= \exp[(100 - 50)(.0204)] \\ &= 2.77. \end{aligned}$$

2. For $X = \log_{10}(\text{Acid})$, we find

$$\hat{\beta}_1 = 5.1683,$$

from which the odds ratio for (Acid = 100) versus (Acid = 50) would be

$$\begin{aligned} \text{OR} &= \exp\{[\log_{10}(100) - \log_{10}(50)][5.1683]\} \\ &= 4.74. \end{aligned}$$

Note: If $X = \text{Acid}$ is used, an SAS program would include these instructions:

```
PROC LOGISTIC DESCENDING;
DATA=CANCER;
MODEL NODES=ACID;
```

where CANCER is the name assigned to the data set, NODES is the variable name for nodal involvement, and ACID the variable name for our covariate, the level of acid phosphatase in blood serum . The option DESCENDING is needed because PROC LOGISTIC models $\Pr(Y = 0)$ instead of $\Pr(Y = 1)$.

The above results are different for two different choices of X and this seems to cause an obvious problem of choosing an appropriate measurement scale. Of course, we assume a *linear* model and one choice of scale for X would fit better than the other. However, it is very difficult to compare different scales unless there were replicated data at each level of X; if such replications are available, one can simply graph a scatter diagram of log(odd) versus X-value and check for linearity of each choice of scale of measurement for X.

4.1.4. Tests of Association

The last two subsections deal with inferences concerning the primary regression coefficient β_1, including both point and interval estimation of this parameter and the odds ratio. Another aspect of statistical inference concerns the test of significance; the null hypothesis to be considered is

$$\mathcal{H}_0 : \beta_1 = 0.$$

The reason for interest in testing whether or not $\beta_1 = 0$ is that $\beta_1 = 0$ implies there is no relation between the binary dependent variable and the covariate X under investigation. Since the likelihood function is rather simple, one can easily derive, say, the score test for the above null hypothesis; however, nothing would be gained by going through this exercise. We can simply apply a chi-square test (if the covariate is binary or categorical) or a t-test or Wilcoxon test (if the covariate under investigation is on a continuous scale). Of course, the application of the logistic model is still desirable, at least in the case of a continuous covariate, because it would provide a measure of association.

4.1.5. The Use of the Logistic Model for Different Designs

Data for risk determination may come from different sources, with the two fundamental designs being retrospective and prospective. Prospective studies enroll groups of subjects and follow them over certain periods of time—examples include occupational mortality studies and clinical trials—and observe the occurrence of certain events of interest such as a disease or death. Retrospective studies gather past data from selected cases and controls to determine differences, if any, in the exposure to a suspected risk factor. They are commonly referred to as case-control studies. It can be seen that the logistic model fits in very well with the prospective or follow-up studies and has been used successfully to model the "risk" of developing a condition—say, a disease—over a specified time period as a function of certain risk factor. In such applications, after a logistic model has been fitted, one can estimate the individual risks $\pi(x)$'s—given the covariate value x—as well as any risk ratio or relative risk,

$$\mathrm{RR} = \frac{\pi(x_i)}{\pi(x_j)}.$$

As for case-control studies, it can be shown—using the so-called Bayes' theorem—that if we have for the population

$$\Pr(Y = 1;\ \text{given } x) = \frac{1}{1 + \exp[-(\beta_0 + \beta_1 x)]},$$

then

$$\Pr(Y = 1;\ \text{given the samples and given } x) = \frac{1}{1 + \exp[-(\beta_0^* + \beta_1 x)]}$$

with

$$\beta_0^* = \beta_0 + \frac{\theta_1}{\theta_0},$$

where θ_1 is the probability that a case was sampled and θ_0 is the probability that a control was sampled. This result indicates that, for a case-control study,

(i) We *cannot* estimate individual risks, nor relative risk, unless θ_0 and θ_1 are known, which are unlikely. The value of the "intercept" provided by the computer output is meaningless.

(ii) However, since we have the same β_1 as with the population model, we still can estimate the odds ratio and, if the rare disease assumption applies, can interpret the numerical result as an approximate relative risk.

4.1.6. Overdispersion

Logistic regression is based on the point-binomial or Bernouilli distribution; its mean is π and variance is $(\pi)(1-\pi)$. If the true variance is greater than $\pi(1-\pi)$ then we have an overdispersed model. Overdispersion is a common phenomenon in practice and it causes concerns because the implication is serious; the analysis which assumes the logistic model often underestimates standard error(s) and, thus, wrongly inflates the level of significance.

Measuring and Monitoring Dispersion

After a logistic regression model is fitted, dispersion is measured by the scaled deviance or scaled Pearson chi-square; it is the deviance or Pearson chi-square divided by the degrees of freedom. The deviance is defined as twice the difference between the maximum achievable log-likelihood and the log-likelihood at the maximum likelihood estimates of the regression parameters. Suppose that data are with replications consisting of m subgroups (with identical covariate values), then the Pearson chi-square and deviance are given by:

$$X_P^2 = \sum_i \frac{[r_i - n_i p_i]^2}{n_i p_i}$$

$$X_D^2 = \sum_i r_i \log\left[\frac{r_i}{n_i p_i}\right].$$

Each of these goodness-of-fit statistics devided by the appropriate degrees of freedom, called the scaled Pearson chi-square and scaled deviance, respectively, can be used as a measure for overdispersion (underdispersion, with those measures less than one, occurs much less often in practice). When their values are much larger than one, the assumption of binomial variability may not be valid and the data are said to exhibit overdispersion. Several reasons can cause overdispersion; among these are such problems as outliers in the data, omitting important covariates in the model, and the need to transform some explanatory factors.

The SAS program PROC LOGISTIC has an option, called AGGREGATE, that can be used to form subgroups. Without such grouping, data may be too sparse, the Pearson chi-square and deviance do not have a chi-square distribution, and the scaled Pearson chi-square and scaled deviance cannot be used as indicators of overdispersion. A large difference between the scaled Pearson chi-square and scaled deviance provides evidence of this situation.

Fitting an Overdispersed Logistic Model

One way of correcting overdispersion is to multiply the covariance matrix by the value of the overdispersion parameter ϕ, the scaled Pearson chi-square, or scaled deviance (as used in weighted least square fitting):

$$E(p_i) = \pi_i$$

$$\mathrm{Var}(p_i) = \phi\pi_i(1 - \pi_i).$$

In this correction process, the parameter estimates are not changed. However, their standard errors are adjusted (increased), affecting their significant levels (reduced).

Example 4.3. In a study of the toxicity of a certain chemical compound, five groups of 20 rats each were fed for four weeks with a diet mixed with that compound at five different doses. At the end of the study, their lungs were harvested and subjected to histopathological examinations to observe for sign(s) of toxicity (yes = 1, no = 0). The results were:

Group	Dose (mg)	Number of Rats	No. of Rats with Toxicity
1	5	20	1
2	10	20	3
3	15	20	7
4	20	20	14
5	30	20	10

A routine fit of the simple logistic regression model yields:

Variable	Coefficient	St. Error	z Statistic	p-Value
Intercept	−2.3407	0.5380	−4.3507	0.0001
Dose	0.1017	0.0277	3.6715	0.0002

In addition, we obtained these results for the monitoring of overdispersion:

Parameter	Chi-Square	Degrees of Freedom	Scaled Parameter
Pearson	10.9919	3	3.664
Deviance	10.7863	3	3.595

Note: An SAS program would include these instructions:

```
INPUT DOSE N TOXIC;
PROC LOGISTIC DESCENDING;
MODEL TOXIC/N= DOSE/SCALE=NONE;
```

The above results indicate an obvious sign of overdispersion. By fitting an overdispersed model, controlling for the scaled deviance, we have:

Variable	Coefficient	St. Error	z Statistic	p-Value
Intercept	−2.3407	1.0297	−2.2732	0.0230
Dose	0.1017	0.0530	1.9189	0.0548

As compared to the previous results, the point estimates remain the same but the standard errors are larger; the effect of Dose is no longer significant at the 5 percent level.

Note: An SAS program would include these instructions:

```
INPUT DOSE N TOXIC;
PROC LOGISTIC DESCENDING;
MODEL TOXIC/N= DOSE/SCALE=D;
```

4.2. MULTIPLE REGRESSION ANALYSIS

The effect of some factor on a dependent or response variable may be influenced by the presence of other factors through effect modifications (i.e., interactions). Therefore, in order to provide a more comprehensive analysis, it is very desirable to consider a large number of factors and sort out which ones are most closely related to the dependent variable. In this section, we will discuss a multivariate method for risk determination. This method, which is multiple logistic regression analysis, involves a linear combination of the explanatory or independent variables; the variables must be quantitative with particular numerical values for each patient. A covariate or independent variable—such as a patient characteristic—may be dichotomous, polytomous, or continuous (categorical factors will be represented by dummy variables). Examples of dichotomous covariates are sex, and presence or absence of certain co-morbidity. Polytomous covariates include race, and different grades of symptoms; these can be covered by the use of *dummy* variables. Continuous covariates include patient age, blood pressure, etc.; In many cases, data transformations (e.g., taking the logarithm) may be desirable to satisfy the linearity assumption.

4.2.1. Logistic Regression Model with Several Covariates

Suppose we want to consider k covariates simultaneously; the simple logistic model of the previous section can be easily generalized and expressed as

$$\pi_i = \frac{1}{1 + \exp[-(\beta_0 + \sum_{j=1}^{k} \beta_j x_{ji})]}, \qquad i = 1,2,\ldots,n,$$

or, equivalently,

$$\ln \frac{\pi_i}{1 - \pi_i} = \beta_0 + \sum_{j=1}^{k} \beta_j x_{ji}.$$

This leads to the likelihood function

$$L = \prod_{i=1}^{n} \frac{\{\exp[\beta_0 + \sum_{i=1}^{k} \beta_j x_{ji}]\}^{y_i}}{1 + \exp[\beta_0 + \sum_{j=1}^{k} \beta_j x_{ji}]}; \qquad y_i = 0, 1,$$

from which parameters can be estimated iteratively using a computer packaged program such as SAS.

Also similar to the univariate case, $\exp(\beta_i)$ represents:

(i) the odds ratio associated with an exposure if X_i is binary (exposed $X_i = 1$ vs. unexposed $X_i = 0$), or

(ii) the odds ratio due to one unit increase if X_i is continuous ($X_i = x + 1$ vs. $X_i = x$).

After $\hat{\beta}_i$ and its standard error have been obtained, a 95 percent confidence interval for the above odds ratio is given by:

$$\exp[\hat{\beta}_i \pm 1.96\mathrm{SE}(\hat{\beta}_i)].$$

These results are necessary in the effort to identify important risk factors for the binary outcome. Of course, before such analyses are done, the problem and the data have to be examined carefully. If some of the variables are highly correlated, then one or fewer of the correlated factors are likely to be as good predictors as all of them; information from other similar studies also has to be incorporated so as to drop some of these correlated explanatory variables. The uses of products, such as $X_1 X_2$, and higher power terms, such as X_1^2, may be necessary and can improve the goodness-of-fit. It is important to note that we are assuming a *(log)linear* regression model in which, for example, the odds ratio due to one unit increase in the value of a continuous X_i ($X_i = x + 1$ vs. $X_i = x$) is independent of x. Therefore, if this *linearity* seems to be violated, the incorporation of powers of X_i should be seriously considered. The use of products will help in the investigation of possible effect modifications. And, finally, there is the messy problem of missing data. Most packaged programs would delete a subject if one or more covariate values are missing.

4.2.2. Effect Modifications

Consider the model:

$$\pi_i = \frac{1}{1 + \exp[-(\beta_0 + \beta_1 x_{1i} + \beta_2 x_{2i} + \beta_3 x_{1i} x_{2i})]},$$
$$i = 1,2,\ldots,n.$$

The meaning of β_1 and β_2 here is not the same as that given earlier because of the cross-product term $\beta_3 x_1 x_2$. Suppose that both X_1 and X_2 are binary; then:

1. for $X_2 = 1$ or exposed, we have

$$(\text{odds; not exposed to } X_1) = e^{\beta_2}$$
$$(\text{odds; exposed to } X_1) = e^{\beta_1 + \beta_2 + \beta_3},$$

   so that the ratio of these odds, $e^{\beta_1 + \beta_3}$, represents the odds ratio associated with X_1, exposed versus nonexposed, in the presence of X_2, whereas

2. for $X_2 = 0$ or not exposed, we have

$$(\text{odds; not exposed to } X_1) = 1.0 \text{ (i.e., baseline)}$$
$$(\text{odds; exposed to } X_1) = e^{\beta_1},$$

   so that the ratio of these odds, e^{β_1}, represents the odds ratio associated with X_1, exposed versus nonexposed, in the absence of X_2.

In other words, the effect of X_1 depends on the level (presence or absence) of X_2 and vice versa. This phenomenon is called *effect modification*, (i.e., one factor modifies the effect of the other). The cross-product term $x_1 x_2$ is called an interaction term. The use of these products will help in the investigation of possible effect modifications. If $\beta_3 = 0$, the effect of two factors acting together, as measured by the odds ratio, is equal to the combined effects of two factors acting separately, as measured by the product of two odds ratios:

$$e^{\beta_1+\beta_2} = e^{\beta_1} \cdot e^{\beta_2}.$$

This fits the classic definition of *no interaction* on a multiplicative scale.

4.2.3. Polynomial Regression

Consider the model,

$$\pi_i = \frac{1}{1 + \exp[-(\beta_0 + \beta_1 x_i + \beta_2 x_i^2)]}, \qquad i = 1,2,\ldots,n,$$

where X is a continuous covariate. The meaning of β_1 here is not the same as that given earlier because of the quadratic term $\beta_2 x_i^2$. We have, for example,

$$\ln(\text{Odd};\ X = x) = \beta_0 + \beta_1 x + \beta_2 x^2$$

$$\ln(\text{Odd};\ X = x + 1) = \beta_0 + \beta_1(x + 1) + \beta_2(x + 1)^2,$$

so that the difference leads, after exponentiating, to

$$\begin{aligned} \text{OR} &= \frac{(\text{Odd};\ X = x + 1)}{(\text{Odd};\ X = x)} \\ &= \exp[\beta_1 + \beta_2(2x + 1)], \end{aligned}$$

a function of x.

Polynomial models with an independent variable present in higher powers than the second are not often used. The second-order or quadratic model has two basic type of uses: (i) when the true relationship is a second-degree polynomial or when the true relationship is unknown but the second-degree polynomial provides a better fit than a linear one, but (ii) more often, a quadratic model is fitted for the purpose of establishing the linearity. The key item to look for is whether $\beta_2 = 0$.

The use of polynomial models is not without drawbacks. The most potential drawback is that multicolinearity is unavoidable; especially, if the covariate is restricted to a narrow range, then the degree of multicolinearity can be quite high. Another problem arises when one wants to use the stepwise regression search method. In addition, finding a satisfactory interpretation for the *curvature* effect coefficient β_2 is not easy.

4.2.4. Testing Hypotheses in Multiple Logistic Regression

Once we have fit a multiple logistic regression model and obtained estimates for the various parameters of interest, we want to answer questions about the contributions of various factors to the prediction of the binary response variable. There are three types of such questions:

- **(i)** An overall test: Taken collectively, does the entire set of explanatory or independent variables contribute significantly to the prediction of response?
- **(ii)** Test for the value of a single factor: Does the addition of one particular variable of interest add significantly to the prediction of response over and above that achieved by other independent variables?
- **(iii)** Test for contribution of a group of variables: Does the addition of a group of variables add significantly to the prediction of response over and above that achieved by other independent variables?

Overall Regression Tests

We now consider the first question stated above concerning an overall test for a model containing k factors, say,

$$\pi_i = \frac{1}{1 + \exp[-(\beta_0 + \sum_{j=1}^{k} \beta_j x_{ji})]}, \qquad i = 1, 2, \ldots, n.$$

The null hypothesis for this test may stated as: "All k independent variables *considered together* do not explain the variation in the responses." In other words,

$$\mathcal{H}_0 : \beta_1 = \beta_2 = \cdots = \beta_k = 0.$$

Two likelihood-based statistics can be used to test this *global* null hypothesis; each has an asymptotic chi-squared distribution with k degrees of freedom under $\mathcal{H}_0$. Both statistics are provided by most standard computer programs such as SAS and they are asymptotically equivalent, yielding identical statistical decisions most of the time.

(i) Likelihood ratio test:

$$\chi^2_{\text{LR}} = 2[\ln L(\hat{\beta}) - \ln L(\mathbf{0})].$$

(ii) Score test:

$$\chi^2_S = \left[\frac{\delta \ln L(\mathbf{0})}{\delta\beta}\right]\left[-\frac{\delta^2 \ln L(\mathbf{0})}{\delta\beta^2}\right]^{-1}\left[\frac{\delta \ln L(\mathbf{0})}{\delta\beta}\right].$$

Example 4.4. Refer to the data set on prostate cancer of Example 4.1 with all five covariates; we have the following test statistics for the global null hypothesis:

(i) Likelihood test:

$$\chi^2_{\text{LR}} = 22.126 \quad \text{with} \quad 5 \text{ dfs}; \qquad p = .0005.$$

(ii) Score test:

$$\chi^2_S = 19.451 \quad \text{with} \quad 5 \text{ dfs}; \qquad p = .0016.$$

Note: An SAS program would include these instructions:

```
PROC LOGISTIC DESCENDING
DATA=CANCER;
MODEL NODES=XRAY GRADE STAGE AGE ACID;
```

where CANCER is the name assigned to the data set, NODES is the variable name for nodal involvement, and XRAY, GRADE, STAGE, AGE, and ACID are the variable names assigned to the five covariates.

Tests for a Single Variable

Let us assume that we now wish to test whether the addition of one particular independent variable of interest adds significantly to the prediction of the response over and above that achieved by other factors already present in the model. The null hypothesis for this test may stated as: "Factor X_i does not have any value added to the prediction of the response *given that other factors are already included in the model.*" In other words,

$$\mathcal{H}_0 : \beta_i = 0.$$

To test such a null hypothesis, one can perform a likelihood ratio chi-squared test, with 1 df, similar to that for the above global hypothesis:

$$\chi^2_{\mathrm{LR}} = 2[\ln L(\hat{\beta};\ \text{all } X\text{'s}) - \ln L(\hat{\beta};\ \text{all other } X\text{'s with } X_i \text{ deleted})].$$

A much easier alternative method is using

$$z_i = \frac{\hat{\beta}_i}{\mathrm{SE}(\hat{\beta}_i)},$$

where $\hat{\beta}_i$ is the corresponding estimated regression coefficient and $\mathrm{SE}(\hat{\beta}_i)$ is the estimate of the standard error of $\hat{\beta}_i$, both of which are printed by standard computer packaged programs. In performing this test, we refer the value of the z statistic to percentiles of the standard normal distribution.

Example 4.5. Refer to the data set on prostate cancer of Example 4.1 with all five covariates; we have:

Variable	Coefficient	St. Error	z Statistic	p-Value
Intercept	0.0618	3.4599	0.018	0.9857
Xray	2.0453	0.8072	2.534	0.0113
Stage	1.5641	0.7740	2.021	0.0433
Grade	0.7614	0.7708	0.988	0.3232
Age	−0.0693	0.0579	−1.197	0.2314
Acid	0.0243	0.0132	1.850	0.0643

The effects of Xray and Stage are significant at the 5 percent level whereas the effect of Acid is marginally significant ($p = .0643$).

Note: Use the same SAS program as in the previous example.

Given a continuous variable of interest, one can fit a polynomial model and use this type of test to check for linearity. It can also be used to check for a single product representing an effect modification.

Example 4.6. Refer to the data set on prostate cancer of Example 4.1, but this time we investigate only one covariate, the level of acid phosphatase (Acid). After fitting the second-degree polynomial model,

$$\pi_i = \frac{1}{1 + \exp[-(\beta_0 + \beta_1(\text{Acid}) + \beta_2(\text{Acid})^2)]}, \qquad i = 1, 2, \ldots, n,$$

we obtained the following results indicating that the *curvature effect* should not be ignored ($p = 0.0437$):

Factor	Coefficient	St. Error	z Statistic	p-Value
Intercept	−7.3200	2.6229	−2.791	0.0053
Acid	0.1489	0.0609	2.445	0.0145
Acid^2	−0.0007	0.0003	−2.017	0.0437

Contribution of a Group of Variables

This testing procedure addresses the more general problem of assessing the additional contribution of two or more factors to the prediction of the response over and above that made by other variables already in the regression model. In other words, the null hypothesis is of the form

$$\mathcal{H}_0 : \beta_1 = \beta_2 = \cdots = \beta_m = 0.$$

To test such a null hypothesis, one can perform a likelihood ratio chi-squared test, with m df,

$$\chi^2_{\text{LR}} = 2[\ln L(\hat{\beta}; \text{ all } X\text{'s}) - \ln L(\hat{\beta}; \text{ all other } X\text{'s with } X\text{'s under investigation deleted})]$$

As with the above *z test*, this *multiple contribution* procedure is very useful for assessing the importance of potential explanatory variables. In particular it is often used to test whether a similar group of variables, such as *demographic characteristics*, is important for the prediction of the response; these variables have some trait in common. Another application would be a collection of powers and/or product terms (referred to as interaction variables). It is often of interest to assess the interaction effects collectively before trying to consider individual interaction terms in a model as previously suggested. In fact, such use reduces the total number of tests to be performed and this, in turn, helps to provide better control of overall Type I error rates which may be inflated due to multiple testing.

Example 4.7. Refer to the data set on prostate cancer of Example 4.1 with all five covariates, and we consider, collectively, these four interaction terms: Acid * Xray, Acid * Stage, Acid * Grade, and Acid * Age. The basic idea is to see if *any* of the other variables would modify the effect of the level of acid phosphatase on the response.

1. With the original five variables, we obtained: $\ln L = -24.063$.
2. With all nine variables, five original plus four products, we obtained: $\ln L = -20.378$.

Therefore:

$$\begin{aligned}
\chi^2_{\text{LR}} &= 2[\ln L(\hat{\beta};\ \text{nine variables}) \\
&\qquad - \ln L(\hat{\beta};\ \text{five original variables})] \\
&= 7.371; \qquad 4\ \text{dfs}, \qquad .05 \le p\text{-value} \le .10.
\end{aligned}$$

In other words, all four interaction terms, *considered together*, are marginally significant ($.05 \le p$-value $\le .10$); there may be some weak effect modification and the effect of acid phosphatase on the response may be somewhat stronger for a certain combination of levels of the other four variables.

Stepwise Regression

In many applications (e.g., a case-control study on a specific disease), our major interest is to identify important risk factors. In other words, we wish to identify from many available factors a small subset of factors that relate significantly to the outcome (e.g., the disease under investigation). In that identification process, of course, we wish to avoid a large Type I (false positive) error. In a regression analysis, a Type I error corresponds to including a predictor that has no real relationship to the outcome; such an inclusion can greatly confuse the interpretation of the regression results. In a standard multiple regression analysis, this goal can be achieved by using a strategy that adds into or removes from a regression model one factor at a time according to a certain order of relative importance. Therefore, the two important steps are:

1. Specifying a criterion or criteria for selecting a model.
2. Specifying a strategy for applying the chosen criterion or criteria.

Strategies: This is concerned with specifying the strategy for selecting variables. Traditionally, such a strategy is concerned with whether a particular variable should be added to a model or whether any particular variable should be deleted from a model at a particular stage of the process. As computers became more accessible and more powerful, these practices became more popular.

Forward Selection Procedure In the forward selection procedure, we proceed as follows:

Step 1: Fit a simple logistic linear regression model to each factor, one at a time.

Step 2: Select the most important factor according to a certain predetermined criterion.

Step 3: Test for the significance of the factor selected in step 2 and determine, according to a certain predetermined criterion, whether or not to add this factor to the model.

Step 4: Repeat steps 2 and 3 for those variables not yet in the model. At any subsequent step, if none meets the criterion in step 3, no more variables are included in the model and the process is terminated. □

Backward Elimination Procedure In the backward elimination procedure, we proceed as follows:

Step 1: Fit the multiple logistic regression model containing all available independent variables.

Step 2: Select the least important factor according to a certain predetermined criterion; this is done by considering one factor at a time and treating it as though it were the last variable to enter.

Step 3: Test for the significance of the factor selected in step 2 and determine, according to a certain predetermined criterion, whether or not to delete this factor from the model.

Step 4: Repeat steps 2 and 3 for those variables still in the model. At any subsequent step, if none meets the criterion in step 3, no more variables are removed from the model and the process is terminated. □

Stepwise Regression Procedure Stepwise regression is a modified version of forward regression that permits reexamination, at every step, of the variables incorporated in the model in previous steps. A variable entered at an early stage may become superfluous at a later stage because of its relationship with other variables now in the model; the information it provides becomes redundant. That variable may be removed, if meeting the elimination criterion, and the model is re-fitted with the remaining variables, and the forward process goes on. The whole process, one step forward followed by one step backward, continues until no more variables can be added or removed. □

Criteria: For the first step of the forward selection procedure, decisions are based on individual score test results (chi-squared, 1 df). In subsequent steps, both forward and backward, the ordering of levels of importance (step 2) and the selection (test in step 3) are based on the likelihood ratio chi-squared statistic:

$$\chi^2_{\mathrm{LR}} = 2[\ln L(\hat{\beta};\ \text{all other } X\text{'s}) - \ln L(\hat{\beta};\ \text{all other } X\text{'s with one } X \text{ deleted})].$$

Example 4.8. Refer to the data set on prostate cancer of Example 4.1 with all five covariates: Xray, Stage, Grade, Age, and Acid. This time we perform a stepwise regression analysis in which we specify that a variable has to be significant at the 0.10 level before it can enter into the model and that a variable in the model has to be significant at the 0.15 for it to remain in the model (most standard computer programs allow users to make these selections; default values are available). First, we get these individual score test results for all variables:

Variable	Score χ^2	p-Value
Xray	11.2831	0.0008
Stage	7.4383	0.0064
Grade	4.0746	0.0435
Age	1.0936	0.2957
Acid	3.1172	0.0775

These indicate that Xray is the most significant variable; thus:

Step 1: Variable Xray is entered.

Analysis of Variables Not in the Model:

Variable	Score χ^2	p-value
Stage	5.6394	0.0176
Grade	2.3710	0.1236
Age	1.3523	0.2449
Acid	2.0733	0.1499

Step 2: Variable Stage is entered.

Analysis of Variables in the Model:

Factor	Coefficient	St. Error	z Statistic	p-Value
Intercept	−2.0446	0.6100	−3.352	0.0008
Xray	2.1194	0.7468	2.838	0.0045
Stage	1.5883	0.7000	2.269	0.0233

Neither variable is removed.

Analysis of Variables Not in the Model:

Variable	Score χ^2	p-Value
Grade	0.5839	0.4448
Age	1.2678	0.2602
Acid	3.0917	0.0787

Step 3: Variable Acid is entered.

Analysis of Variables in the Model:

Factor	Coefficient	St. Error	z Statistic	p-Value
Intercept	−3.5756	1.1812	−3.027	0.0025
Xray	2.0618	0.7777	2.651	0.0080
Stage	1.7556	0.7391	2.375	0.0175
Acid	0.0206	0.0126	1.631	0.1029

None of the variables are removed.

Analysis of Variables Not in the Model:

Variable	Score χ^2	p-Value
Grade	1.065	0.3020
Age	1.5549	0.2124

No (additional) variables meet the 0.1 level for entry into the model.

Note: An SAS program would include these instructions:

```
PROC LOGISTIC DESCENDING
DATA=CANCER;
MODEL NODES=XRAY, GRADE, STAGE, AGE, ACID
/SELECTION=STEPWISE SLE=.10 SLS=.15 DETAILS;
```

where CANCER is the name assigned to the data set, NODES is the variable name for nodal involvement, and XRAY, GRADE, STAGE, AGE, and ACID are the variable names assigned to the five covariates. The option DETAILS provides step-by-step detailed results; without specifying it, we would have only the final fitted model (which is just fine in practical applications). The default values for SLE (entry) and SLS (stay) probabilities are .05 and .10, respectively.

4.2.5. The Receiver Operating Characteristic (ROC) Curve

Screening tests, as presented in section 2.1, were focused on binary test outcome. However, it is often true that the result of the test, although dichotomous, is based on the dichotomization of a continuous variable—say, X—herein referred to as the *separator variable.* Let us assume without loss of generality that smaller values of X are associated with the diseased population, often called the *population of the cases.* Conversely, larger values of the separator are assumed to be associated with the control or nondiseased population.

A test result is classified by choosing a cutoff $X = x$ against which the observation of the separator is compared. A test result is positive if the value of the separator does not exceed the cutoff; otherwise, the result is classified as negative. Most diagnostic tests are imperfect instruments in the sense that healthy individuals will occasionally be classified wrongly as being ill, while some individuals who are really ill may fail to be detected as such. Therefore, there is the ensuing conditional probability of the correct classification of a randomly selected case, the sensitivity of the test defined as in section 2.1:

$$F(x) = \Pr(X \leq x \mid \text{case}).$$

This is the *cumulative distribution function* (cdf) of the separator variable for the cases. Similarly, the conditional probability of the correct classification of a randomly selected control, $\Pr(X > x \mid \text{control})$, is the specificity of the test. It is obvious that

$$\begin{aligned} G(x) &= 1 - \text{specificity} \\ &= \Pr(X \leq x \mid \text{control}) \end{aligned}$$

is the cumulative distribution function (cdf) of the separator variable for the controls.

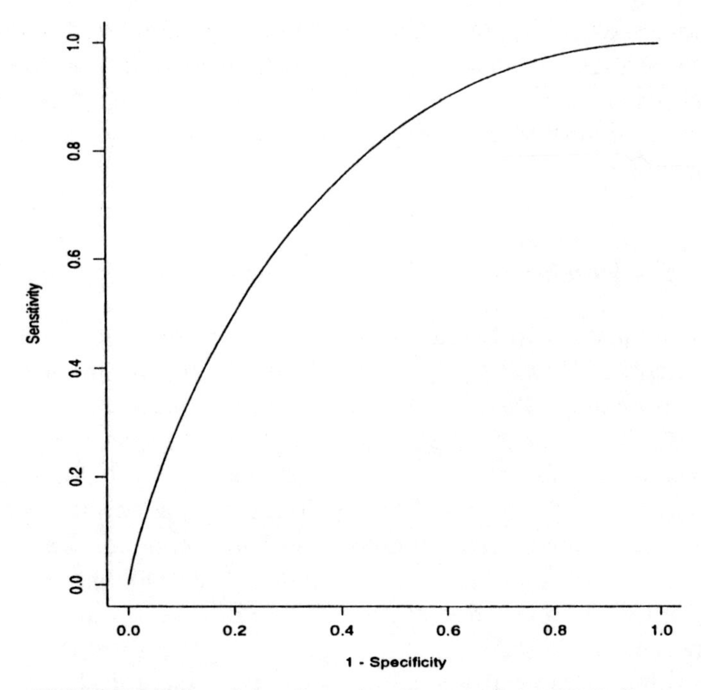

FIGURE 4.2. Receiving Operating Characteristic (ROC) Curve

A *receiver operating characteristic* (ROC) curve (Figure 4.2), the trace of the sensitivity versus (1 – specificity) of the test, is generated as the cutoff x moves through its range of possible values. In other words, the ROC function R associated with a separator X is defined by

$$F(x) = R[G(x)].$$

Given two independent samples, $\{x_{1i}\}_{i=1}^{m}$ and $\{x_{2j}\}_{j=1}^{n}$, from m controls and n cases respectively, the cdfs F and G are estimated by the empirical cdfs defined by

$$F_n(x) = (\text{number of } x_{2j}\text{'s} \leq x)/n$$
$$G_m(x) = (\text{number of } x_{1i}\text{'s} \leq x)/m.$$

This estimator $\hat{R}$, often called the nonparametric ROC curve, is defined by

$$F_n(x) = \hat{R}[G_m(x)]$$

(Bamber, 1975; Hanley and McNeil, 1982). Graphically, $\hat{R}$ is a random walk from left-bottom corner (0,0) to right-top corner (1,1) whose next step is $1/m$ to the right or $1/n$ up according to whether the next observation in the ordered combined sample is a control (x_1) or a case (x_2). For example, suppose we have the samples

$$x_{21} < x_{22} < x_{11} < x_{23} < x_{12} \qquad (n = 3,\ m = 2);$$

then $\hat{R}$ is the following step function:

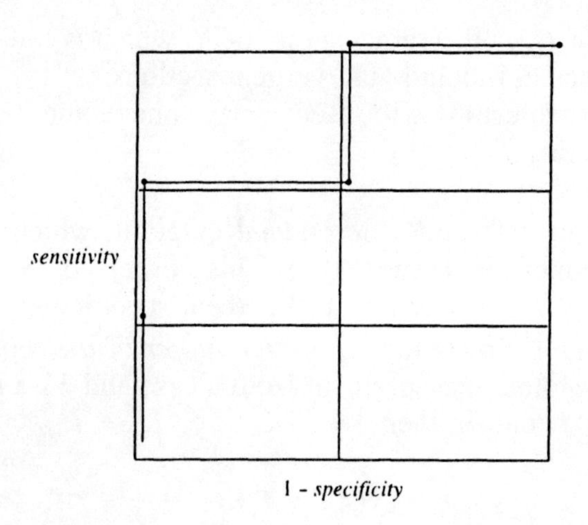

The estimator $\hat{R}$ can also be equivalently expressed as follows (Le, 1997b): Let R_j be the rank of x_{2j} among the cases $\{x_{2j}\}_{j=1}^{n}$ ($1 \le R_j \le n$); S_j be the rank of x_{2j} among the pooled sample $\{x_{1i}, x_{2j}\}$ ($1 \le S_j \le n + m$); and

$$u_j = \frac{S_j - R_j}{m}, \qquad 1 \le j \le n,$$

the percent of controls to the left of case j. Then the nonparametric ROC curve $\hat{R}$ is the step function defined, with the use of mid-ranks or average ranks for tied data, by

$$\hat{R}(u) = (\text{number of } u_j \le u)/n, \qquad 0 \le u \le 1.$$

It is a nonparametric estimator depending on the ranks of the measurements only. For example, suppose we have the same samples as above,

$$x_{21} < x_{22} < x_{11} < x_{23} < x_{12} \qquad (n = 3,\ m = 2);$$

then R_3 is a step function defined by

$$R_3(y) = \begin{cases} 2/3 & \text{for} \quad 0 \le y < 1/2 \\ 1 & \text{for} \quad 1/2 \le y \le 1. \end{cases}$$

The graph of R_3 is the same as that of $\hat{R}$, which is obtainable using SAS, but does not include the vertical sections.

Being able to estimate the ROC curve, one would be able to do a number of things:

(i) We can determine the *optimal* cutpoint, which is nearest to the upper-left corner (0,1). This corner corresponds to 100 percent sensitivity and 100 percent specificity.

(ii) We can estimate the *separation power* of the separator X. Let X_2 denote a measurement from a case and X_1 a measurement from a control, then,

$$\begin{aligned} \theta &= \Pr(X_2 < X_1) \\ &= \int F\,dG \\ &= \int_0^1 R(x)\,dx \\ &= \text{Area under the ROC curve.} \end{aligned}$$

Using the above nonparametric ROC curve $\hat{R}$, the estimated separation power $\hat{\theta}$ is only a function of the Wilcoxon rank-sum statistic $\sum S_j$,

$$\hat{\theta} = \frac{\Sigma(1 - u_j)}{n}.$$

Given two available separators, the better separator is the one with higher separation power.

In recent years, diagnostic research has begun to address the potentialities and the problems associated with the incorporation of concomitant information into *receiver operating characteristic* (ROC) studies. For example, Ahlquist et al. (1985), when studying the diagnostic performance of a dye test for the detection of fecal blood, found that its specificity and sensitivity were markedly affected by the condition of the sample material itself (e.g., wetness, temperature, etc.). Similarly, when studying the diagnostic performances of admittance and tympanometric width for the detection of middle ear fluid (Le et al., 1992), we found that their diagnostic powers were affected by otoscopic findings. Hence, researchers often look for methods for the inclusion of concomitant information when evaluating diagnostic tests.

A natural approach to account for extraneous factors is regression methodology. An example of such work is that of Hlatky et al. (1984), who employed multivariate logistic regression to explore factors influencing the sensitivity and specificity of exercise electrocardiography. In a similar effort, Tosteson and Begg (1988) investigated the application of generalized ordinal regression models for the inclusion of covariates into the analysis of ratings-based ROC curves. Basically, if $X_1, X_2, \ldots, X_k$ denote the k confounders, would would investigate the effects of $XX_1, XX_2, \ldots, XX_k$ in a multiple logistic regression model. The regression coefficient of the product term XX_i represents the confounding effect of X_i.

4.2.6. Measures of Goodness-of-Fit

In usual (i.e., Gaussian) regression analyses, R^2 gives the proportional reduction in variation in comparing the conditional variation of the response with the marginal variation. It describes the strength of the association between the response and the set of independent variables considered together; for example, with $R^2 = 1$ we can predict the response perfectly. We present here three measures of goodness-of-fit for the logistic regression model; however, none has all the advantages of the coefficient of determination, R^2.

(i) The first option is a likelihood-based measure proposed by McFadden (1974), defined by

$$D = \frac{\ln L(\mathbf{0}) - \ln L(\hat{\beta})}{\ln L(\mathbf{0})},$$

where $\ln L(\hat{\beta})$ denote the maximized log likelihood for the fitted model and $\ln L(\mathbf{0})$ the maximized log likelihood for the "null" model containing only an intercept term. Both of these numbers, $\ln L(\hat{\beta})$ and $\ln L(\mathbf{0})$, are provided by most computer packaged program such as SAS.

(ii) The second measure is a member of a family developed by Efron (1978). Basically, this measure compares the binary reponses y_i's and their corresponding fitted probability values $\hat{\pi}_i'$'s. It represents the proportional reduction in error obtained by using $\hat{\pi}_i$ instead of $\bar{y}$ as a predictor of y_i where $\bar{y}$ is the overall proportion of the sample with $y = 1$. This gives

$$E = 1 - \frac{\sum(y_i - \hat{\pi}_i)^2}{\sum(y_i - \bar{y})^2}.$$

This and the previous measure, D, are not easy to interpret.

(iii) The last measure we present is obtained by an application of a receiver-operator characteristic curve (ROC) analysis of the previous subsection. After fitting a logistic regression model, each subject's fitted response probability, $\hat{\pi}_i$, is calculated as in the previous measure, E. Using these probabilities *as values of a separator*, we can construct a nonparametric ROC curve tracing sensitivities against the estimated false positivities for various cutpoints. Such an ROC curve not only makes it easy to determine an optimal cutpoint (the point on the curve nearest to the top-left corner $(0, 1)$ which corresponds to 1.0 sensitivity and 1.0 specificity) but it also shows the overall performance of the fitted logistic regression model; the better the performance the further the curve is away from the diagonal. The area, C, under this ROC curve can be used as a measure of goodness-of-fit. Measure C represents the *separation power* of the logistic model under consideration; for example, with $C = 1$, the fitted response probabilities for subjects with $y = 1$ and the fitted response probabilities for subjects with $y = 0$ are completely *separated.* A function of C can also be used; for example,

$$C^* = \frac{A - .5}{.5}$$

may have a more desirable range, between 0 and 1. To minimize the risk of optimistic bias that occurs when a fitted model is applied to the same data to which it was fit, a *jackknife* approach should be used. In this approach, each subject's fitted response probability was calculated from a regression analysis generated by excluding his or her data when fitting the model. Of course, the computing is very tedious and there are no computer packaged programs for the task (it is possible to use PROC LOGISTIC of SAS with the addition of a macro file).

Example 4.9. Refer to the data set on prostate cancer of Example 4.1 with all five covariates and fitted results shown in Example 4.4. We have $\ln L(\mathbf{0}) = -35.126$ and $\ln L(\hat{\beta}) = -24.063$, leading to:

$$D = \frac{\ln L(\mathbf{0}) - \ln L(\hat{\beta})}{\ln L(\mathbf{0})}$$
$$= .315.$$

Using the estimated regression parameters obtained from Example 4.4 (where the model was fitted once), we have $E = .362$ and $C = .845$ ($C^* = .690$). If the jackknife approach is applied, then the results would be $E = .187$ and $C = .742$, representing some expected decreases (measure C is the most stable under this kind of cross-validation).

Note: The area under the ROC curve (i.e., measure C), is provided by SAS's PROC LOGISTIC. The computer output also includes another measure by Cox and Snell (1989).

Since the measure of goodness-of-fit A has a meaningful interpretation and increases when we add an explanatory variable to the model, it can be used as a criterion in performing stepwise logistic regression instead of a p-value which is easily influenced by the sample size. For example, in the forward selection procedure, we proceed as follows:

Step 1: Fit a simple logistic linear regression model to each factor, one at a time.

Step 2: Select the most important factor defined as the one with the largest value of the measure of goodness-of-fit A.

Step 3: Compare this value of A for the factor selected in step 2 and determine, according a predetermined criterion, whether or not to add this factor to the model—say, to see if $A \geq .53$, an increase of .03 or 3 percent over .5 when no factor is considered.

Step 4: Repeat steps 2 and 3 for those variables not yet in the model. At any subsequent step, if none meets the criterion in step 3—say, increase the separation power by .03, no more variables are included in the model and the process is terminated.

4.3. ORDINAL LOGISTIC MODEL

As seen in the previous two sections, logistic regression is most often used for a dichotomous response where it models the *logit* of the probability π of having an "event" (for example, a disease), where logit(π) is defined by

$$\text{logit}(\pi) = \log\left(\frac{\pi}{1-\pi}\right),$$

the "log odd" of an event. Suppose we want to consider k covariates simultaneously; the logistic model is expressed as:

$$\text{logit}(\pi) = \beta_0 + \sum_{j=1}^{k} \beta_j x_{ji}.$$

This common logistic model is now generalized for use with an ordinal response having more than two levels. Such a targeted ordinal response has several categories with some natural ordering but may be without a defined metric (i.e., no numeric value associated with each category). For example, the *condition* of a disease can be classified as "normal, mild, moderate, and severe," a four-level ordinal response. For any ordinal response, we can defined *cumulative logits*. For example, if the ordinal response under investigation has k levels with corresponding polynomial probabilities $\pi_1, \pi_2, \ldots, \pi_k$, where

$$\sum_{i=1}^{k} \pi_i = 1,$$

then the cumulative logits are defined as:

$$F_i = \frac{\sum_{j=1}^{i} \pi_j}{\sum_{j=i+1}^{k} \pi_j}.$$

An ordinal logistic regression model describes a relationship between an ordinal response and a set of explanatory or independent variables. Suppose we want to consider k covariates simultaneously; the *proportional odds model* with m covariates by McCullagh (1980) assumes that

$$\log(F_i) = \beta_{0i} + \sum_{j=1}^{m} \beta_j x_j.$$

In other words, the proportional odds model assumes that each logit follows a linear model which has a separate intercept parameter but other regression parameters (i.e., the *slopes* relating the response to each covariate) are constant across all cumulative logits for different levels of the response. This is similar to the parallelism assumption in bioassay and the proportional hazards model in survival analysis. In addition, since

$$F_i \leq F_{i+1},$$

it follows that there is a constraint on the intercepts:

$$\beta_{0i} \leq \beta_{0(i+1)}.$$

Example 4.10. To provide a simple illustration of ordinal logistic regression, let us consider the example in Hanley and McNeil (1983). They studied 112 phantoms that were specially constructed to evaluate the "accuracy" of a computer algorithm used in the image construction for CT. Fifty-eight (58) of these phantoms were of uniform density and were designated as normal, $x = 0$; the remaing 54 contained an area of reduced density to simulate a lesion and were designated as abnormal, $x = 1$. The computer algorithm "reads" each image and rates it on a six-point ordinal scale (Y): 1 = definitely normal, 2 = probably normal, 3 = possibly normal, 4 = possibly abnormal, 5 = probably abnormal, and 6 = definitely abnormal. The results (frequencies) were:

	Computer Reading (Y)					
Status (X)	1	2	3	4	5	6
Normal	12	28	8	6	4	0
Abnormal	1	3	6	13	22	9

Using the computer reading as an ordinal dependent variable, a logistic regression analysis yields:

Factor	Coefficient	St. Error	z Statistic	p-Value
Intercept1	−4.7595	0.5201	−9.151	0.0001
Intercept2	−2.7531	0.3976	−6.924	0.0001
Intercept3	−1.6175	0.3254	−4.970	0.0001
Intercept4	−0.7541	0.2767	−2.725	.0064
Intercept5	1.3146	0.3156	4.165	0.0001
Status(X)	3.0932	0.4464	6.930	0.0001

The results indicate a level of accuracy of the computer algorithm. The odds that an abnormal phantom would be judged as at or above a certain level of abnormality by the computer algorithm, say probably or definitely abnormal (≥ 5), is

$$\begin{aligned} \text{OR} &= e^{3.0932} \\ &= 22.04 \end{aligned}$$

times the odds that a normal phantom would be judged as at or above that same level of abnormality.

Note: An SAS program would include these instructions:

```
DATA PHANTOM;
INPUT EXPLANATORY; RATING COUNT;
CARDS;
```

```
0 1 12
0 2 28
0 3 8
0 4 6
0 5 4
0 6 0
1 1 1
1 2 3
1 3 6
1 4 13
1 5 22
1 6 9
;
PROC LOGISTIC DESCENDING;
WEIGHT COUNT;
MODEL RATING=EXPLANATORY;
```

where EXPLANATORY is the true abnormality (0 being normal), and COUNT is the frequency of each group at each level of reading (RATING) by the the computer algorithm.

We can also obtain almost the same result using the method of Chapter 2. The computer rating would be dichotomized with a moving division point—say, (6) versus (1, 2, 3, 4, 5) then (5, 6) versus (1, 2, 3, 4), etc.—to form five 2×2 tables. We then assume that the odds ratios are constant across tables (the basic proportional odds assumption in the ordinal logistic model); the common odds ratio is then estimated by

$$\text{OR} = \frac{\sum ad}{\sum bc}$$

$$= 22.19.$$

In the following example, taken from Rosner (1982), the response variable has numerical values but we can also treat it as ordinal as well.

Example 4.11. This data set consists of 216 persons aged 20–39 with retinitis pigmentosa (RP). The patients were classified on the basis of a detailed family history into the genetic types of autosomal dominant RP (DOM), autosomal recessive RP (AR), sex-linked RP (SL), and isolate RP (ISO). The *outcome* is obtained from a routine ocular examination; an eye was considered *affected* if visual acuity was 20/50 or worse, and normal otherwise. The response variable (Y) is the number of affected eyes (0, 1, or 2) as tabulated in the following table:

	No. of Eyes		
Type	0	1	2
DOM	15	6	7
AR	7	5	9
SL	3	2	14
ISO	67	24	57

Using the number of affected eyes as an ordinal dependent variable and the autosomal dominant RP (DOM) group as the baseline for comparison (each other group is represented by an indicator variable), a logistic regression analysis yields:

Variable	Coefficient	St. Error	z Statistic	p-Value
Intercept1	−0.9437	0.3726	−2.532	0.0113
Intercept2	−0.2184	0.3671	−0.595	0.5520
AR	0.7688	0.5495	1.399	0.1618
SL	1.9594	0.6323	3.099	0.0019
ISO	0.4392	0.3967	1.107	0.2682

The results indicate that only the SL group is significantly different from the others.

4.4. EXERCISES

Radioactive radon is an inert gas that can migrate from soil and rock and accumulate in enclosed areas such as underground mines and homes. The radioactive decay of trace amounts of uranium in the Earth's crust through radium is the source of radon, or more precisely the isotope radon-222. Radon-222 emits alpha particles; when inhaled, the alpha particles rapidly diffuse across the alveolar membrane of the lung and are transported by the blood to all parts of the body. Due to the relatively high flow rate of blood in bone marrow, this may be a biologically plausible mechanism for the development of leukemia. Table 4.2 provides some data from a case-control study to investigate the association between indoor residential radon exposure and risk of childhood acute myeloid leukemia. The variables are:

- Disease (1 = Case, 2 = Control)
- Radon (radon concentration in Bq/m3)—some chracteristics of the child: Sex (1 = Male, 2 = Female) and Race (1 = White, 2 = Black, 3 = Hispanic, 4 = Asian, and 5 = Others), Downs syndrome (a known risk factor for leukemia; 1 = No, 2 = Yes)
- Risk factors from the parents: Msmoke (1 = Mother a current smoker, 2 = No, 0 = Unknown), Mdrink (1 = Mother a current alcohol drinker, 2 = No, 0 = Unknown), Fsmoke (1 = Father a current smoker, 2 = No, 0 = Unknown), and Fdrink (1 = Father a current alcohol drinker, 2 = No, 0 = Unknown)

  (a) Taken collectively, do the covariates contribute significantly to the separation of the cases and the controls? Give your interpretation for the measure of goodness-of-fit C.

  (b) Fit the multiple regression model to obtain estimates of individual regression coefficients and their standard errors. Draw your conclusion concerning the conditional contribution of each factor.

  (c) Within the context of the multiple regression model in (b), does sex alter the effect of Downs syndrome?

  (d) Within the context of the multiple regression model in (b), does Downs syndrome alter the effect of radon exposure?

  (e) Within the context of the multiple regression model in (b), taken collectively, do smoking-drinking variables (by the father or mother) relate significantly to the disease of the child?

  (f) Within the context of the multiple regression model in (b), is the effect of radon concentration linear?

  (g) Focusing on radon exposure as the primary factor, taken collectively, was this main effect altered by any other covariates?

TABLE 4.2. Radon-Leukemia Data

DISEASE	SEX	RACE	RADON	MSMOKE	MDRINK	FSMOKE	FDRINK	DOWNS
1	2	1	17	1	1	1	2	1
1	2	1	8	1	2	2	0	2
2	2	1	8	1	2	1	2	2
2	1	1	1	2	0	2	0	2
1	1	1	4	2	0	2	0	2
2	1	1	4	1	1	1	1	2
1	2	1	5	2	0	2	0	2
2	1	1	4	2	0	2	0	2
1	2	1	7	1	1	2	0	1
1	1	1	15	1	1	1	2	1
2	1	1	16	2	0	1	1	1
1	1	1	12	2	0	1	2	1
2	1	1	14	1	1	1	1	1
2	2	1	12	2	0	2	0	1
2	1	1	14	1	2	1	2	1
2	2	1	9	1	2	1	1	2
2	1	1	4	2	0	1	1	2
1	1	1	2	2	0	1	1	1
1	2	1	12	2	0	1	1	2
1	2	1	13	2	0	2	0	1
2	2	1	13	2	0	1	2	2
2	1	1	18	2	0	2	0	1
1	1	1	13	1	2	1	1	2
1	2	1	16	2	0	2	0	1
1	2	1	10	1	1	2	0	2
1	1	1	11	2	0	2	0	1
2	2	1	4	1	1	1	1	2
1	2	1	1	1	2	1	1	2
1	1	2	9	2	0	2	0	1
1	2	1	15	1	1	1	2	1
2	2	1	17	2	0	1	2	1
1	1	1	9	2	0	1	2	1
2	1	1	15	2	0	2	0	1
1	1	1	10	1	1	1	1	1
2	1	1	11	1	2	2	0	1
1	2	1	8	2	0	2	0	1
1	1	1	14	1	2	2	0	2
2	2	1	14	1	2	1	2	2
1	2	1	1	2	0	1	1	2
2	1	1	1	2	0	1	1	2
1	2	1	6	1	2	1	2	2
2	1	1	16	2	0	2	0	1
2	2	1	3	2	0	2	0	2
1	2	1	5	2	0	1	2	2
2	1	2	15	2	0	1	9	1
1	1	1	17	2	0	1	2	2
2	1	1	17	1	2	1	2	2
1	2	1	3	2	0	2	0	2
2	1	1	11	2	0	2	0	2

TABLE 4.2. (*Continued*)

DISEASE	SEX	RACE	RADON	MSMOKE	MDRINK	FSMOKE	FDRINK	DOWNS
2	2	1	14	1	2	1	1	2
1	1	1	17	0	1	2	1	1
2	1	1	1	1	2	2	0	2
2	1	1	10	2	0	2	0	2
1	1	1	14	1	1	1	1	1
2	1	1	4	1	2	1	2	2
1	1	3	12	2	0	2	0	2
1	2	1	9	1	1	2	0	2
2	2	1	7	2	0	1	2	2
1	2	1	5	1	2	1	2	2
1	1	1	8	2	0	2	0	2
2	1	1	9	2	0	2	0	2
2	1	1	15	1	2	1	2	2
1	2	1	10	2	0	2	0	1
2	2	1	10	2	0	2	0	1
2	1	1	1	2	0	2	0	2
2	2	1	1	2	0	1	1	2
1	2	1	9	1	1	2	0	1
1	2	5	14	1	1	2	0	2
1	2	1	8	2	0	1	1	2
2	2	1	7	2	0	2	0	2
2	2	1	13	2	0	1	1	2
1	2	1	1	2	0	2	0	2
2	2	1	1	2	0	1	2	2
1	1	5	12	2	0	1	2	1
2	1	1	11	1	2	2	0	1
1	2	1	2	2	0	1	2	2
2	2	1	3	1	1	1	1	2
2	2	1	6	1	2	1	2	1
2	2	1	3	2	0	1	2	2
1	1	3	1	2	0	2	0	2
2	2	5	2	1	2	1	2	2
1	1	5	14	1	1	1	2	1
1	1	1	1	1	1	1	1	2
2	1	1	12	2	0	1	2	1
2	2	1	13	1	1	1	1	2
1	2	1	11	2	0	2	0	1
2	1	1	11	2	0	2	0	1
1	2	1	16	1	2	2	0	1
2	1	1	3	2	0	1	2	2
1	1	1	13	2	0	2	0	1
2	2	1	12	2	0	2	0	1
2	1	1	12	2	0	2	0	1
1	1	1	3	1	2	2	0	2
2	2	1	5	2	0	1	2	2
1	2	1	7	1	1	1	1	1
2	2	1	7	2	0	1	1	1
1	1	1	2	2	0	1	2	2
2	1	1	2	1	1	1	2	2

TABLE 4.2. (*Continued*)

DISEASE	SEX	RACE	RADON	MSMOKE	MDRINK	FSMOKE	FDRINK	DOWNS
1	1	1	2	2	0	1	1	2
2	1	1	2	2	0	2	0	2
2	2	1	2	1	1	1	1	2
1	2	1	3	1	2	2	0	2
2	2	1	3	2	0	2	0	2
2	1	1	14	2	0	2	0	1
2	2	1	15	1	1	1	2	1
1	2	1	1	2	0	1	1	2
2	2	1	1	2	0	1	2	2
2	1	1	1	2	0	1	2	2
2	2	3	10	2	0	1	2	1
2	1	3	9	2	0	1	1	1
2	2	1	14	2	0	2	0	2
2	2	1	9	1	1	1	1	1
2	2	1	9	1	1	1	1	2
1	1	1	17	2	0	1	2	1
2	2	1	3	1	1	1	1	2
2	1	1	5	2	0	1	1	2
1	1	1	15	1	2	1	2	1
2	1	1	14	1	1	2	0	2
2	2	1	5	1	2	1	2	2
2	2	3	13	2	0	1	1	1
2	1	1	15	2	0	1	1	2
2	1	1	12	2	0	1	1	1
1	1	1	1	2	0	1	2	2
2	2	1	2	2	0	2	0	2
2	2	3	1	2	0	1	2	2
2	2	5	3	2	0	2	0	2
2	2	1	15	2	0	2	0	1
1	1	2	6	2	0	2	0	1
2	1	1	5	1	2	1	1	2
1	1	1	2	2	0	2	0	2
1	1	1	15	1	1	2	0	1
1	1	2	5	2	0	1	1	2
1	1	1	2	1	1	1	1	2
2	2	1	2	1	2	1	2	2
1	2	1	17	2	0	1	2	2
2	1	1	8	2	0	2	0	2
2	2	1	6	2	0	1	1	2
1	2	1	1	1	2	1	1	2
1	1	1	13	2	0	1	2	1
2	2	1	12	2	0	1	1	1
2	1	1	9	2	0	2	0	1
1	1	1	10	1	1	2	0	1
2	2	1	15	1	1	2	0	1
1	2	1	3	2	0	1	2	2
2	1	1	2	1	2	1	1	2
1	2	1	11	1	2	1	2	2
2	2	1	11	2	0	1	2	2

TABLE 4.2. (*Continued*)

DISEASE	SEX	RACE	RADON	MSMOKE	MDRINK	FSMOKE	FDRINK	DOWNS
2	1	1	4	2	0	2	0	2
1	2	1	2	2	0	2	0	2
1	2	1	12	2	0	1	1	1
1	2	1	1	2	0	1	2	2
1	1	1	15	1	1	1	2	2
2	2	1	8	2	0	2	0	2
2	2	1	1	1	2	1	1	2
2	2	1	2	1	1	1	1	2
1	1	1	8	1	2	1	1	2
2	1	1	2	2	0	1	2	2
1	1	1	1	1	1	1	1	2
1	2	1	12	2	0	2	0	1
2	1	1	12	1	2	1	1	1
1	2	1	1	2	0	1	1	2
2	1	1	12	2	0	2	0	2
1	2	1	16	2	0	2	0	2
2	2	1	9	2	0	2	0	1
2	2	3	7	2	0	2	0	2
1	2	4	2	2	0	1	2	2
2	1	1	1	2	0	2	0	2
1	2	1	11	1	2	1	2	2
2	2	1	13	1	2	1	2	1
2	2	1	7	1	1	1	1	2
2	1	1	13	1	2	1	2	1
1	2	1	14	1	2	1	2	2
2	1	1	2	2	0	2	0	2
1	2	1	1	1	1	2	0	2
2	2	1	2	2	0	2	0	2
1	2	1	2	2	0	2	0	2
2	2	1	1	1	2	2	0	2
2	1	5	1	2	0	1	1	2
2	1	1	9	1	1	1	2	1
1	1	1	13	1	2	2	0	2
2	1	1	12	2	0	2	0	1
1	2	1	17	2	0	2	0	1
2	2	1	10	2	0	2	0	1
2	1	1	11	1	2	1	2	1
2	2	1	5	2	0	2	0	2
2	2	1	15	2	0	2	0	1
1	2	1	2	2	0	1	1	2
2	2	1	2	1	1	2	0	2
2	2	1	11	1	1	1	1	2
1	1	1	10	2	0	1	9	1
1	1	1	5	1	1	1	2	2
2	1	1	4	1	2	2	0	2
2	2	1	1	1	1	1	1	2
2	2	1	1	2	0	2	0	2
1	2	1	2	2	0	2	0	2
1	2	1	10	1	2	2	0	2

TABLE 4.2. (*Continued*)

DISEASE	SEX	RACE	RADON	MSMOKE	MDRINK	FSMOKE	FDRINK	DOWNS
2	2	1	9	2	0	2	0	1
1	1	1	13	2	0	2	0	2
2	1	1	12	2	0	2	0	1
2	2	1	14	2	0	1	1	2
1	1	1	3	1	2	2	0	2
1	1	5	12	2	0	2	0	2
1	1	1	2	2	0	1	2	2
2	1	1	3	1	1	2	0	2
2	1	1	11	1	1	1	1	2
1	1	1	15	1	1	1	1	2
2	2	1	15	1	1	1	9	1
1	2	1	9	1	2	1	1	1
2	2	1	10	1	1	1	1	1
2	2	1	2	2	0	1	1	2
2	1	1	2	2	0	2	0	2
2	1	1	13	2	0	2	0	2
2	2	1	10	2	0	2	0	2
2	1	1	9	2	0	2	0	2
1	2	1	12	2	0	1	1	1
2	2	1	12	1	2	2	0	2
1	1	1	8	1	2	1	1	1
2	1	1	7	2	0	2	0	2
1	2	1	1	2	0	2	0	2
2	2	1	14	1	2	1	2	1
1	2	1	9	2	0	1	2	2
2	2	1	9	1	2	1	2	1
1	1	5	15	2	0	1	2	1
2	2	1	14	2	0	2	0	2
1	1	1	12	2	0	1	1	1
1	2	1	1	2	0	2	0	2
2	2	1	2	2	0	1	2	2
2	2	3	2	2	0	1	2	2
1	2	1	4	2	0	2	0	2
1	1	1	0	2	0	2	0	2
2	2	1	0	2	0	2	0	2
2	1	1	4	2	0	2	0	2
2	1	1	2	1	2	1	1	2
1	2	3	16	2	0	1	2	2
2	1	1	2	2	0	1	2	2

CHAPTER 5

Methods for Matched Data

Case-control studies have been perhaps the most popular form of research design in epidemiology. They generally can be carried out in a much shorter period of time than cohort studies and are cost effective. As a technique for controlling effects of confounders, randomization and stratification are possible solutions at the design stage and statistical adjustments can be made at the analysis stage. Statistical adjustments are done using regression methods, such as logistic regression in Chapter 4. Stratification is more often introduced at the analysis stage too, and methods such as the Mantel–Haenszel method are available to complete the task. Stratification can also be introduced at the design stage; its advantage is that one can avoid inefficiencies resulting from having some strata with a gross imbalance of cases and controls. A popular form of stratified design occurs when each case is *individually matched* with one or more controls chosen to have similar characteristics (i.e., values of confounding or matching variables). Matched designs have several advantages. They make it possible to control for confounding variables that are difficult to measure directly and, therefore, difficult to adjust at the analysis stage. For example, subject can be matched using area of residence so as to control for environmental exposure. Matching also provides more adequate control of confounding than can adjustment in analysis using regression, because matching does not need any specific assumptions on functional form, which may be needed in regression models. Of course, matching has disadvantages too. Matches for individuals with unusual characteristics are hard to find. In addition, when cases and controls are matched on a certain specific characteristic, the influence of that characterisric on the disease can no longer be studied. Finally, a sample of matched cases and controls is not usually representative of any specific population, which may reduce our ability to generalize the analysis results.

Before coming back to the analysis of matched case-control studies, let's first start with a simple use of matched design to measure agreement.

5.1. MEASURING AGREEMENT

Many research studies rely on an observer's judgment to determine whether a disease, a trait, or an attribute is present or absent. For example, results of ear examinations will surely have effects on a comparison of competing treatments for ear infection. Of course, the basic concern is the issue of reliability. Section 2.1 dealt with an important aspect of reliability, the validity of the assessment. However, in order to judge a method's validity, an exact method for classification, or a *gold standard*, must be available for the calculation of sensitivity and specificity. When an exact method is *not* available, reliability can only be judged *indirectly* in terms of *reproducibility*; the most common way for doing that is measuring the agreement between examiners.

For simplicity, assume that each of two observers independently assigns each of n items or subjects to one of two categories. The sample may then be enumerated in a 2×2 table as follows:

		Observer 2		
		Cat. 1	Cat. 2	Total
Observer 1	Cat. 1	n_{11}	n_{12}	n_{1+}
	Cat. 2	n_{21}	n_{22}	n_{2+}
	Total	n_{+1}	n_{+2}	n

or, in terms of the cell probabilities,

		Observer 2		
		Cat. 1	Cat. 2	Total
Observer 1	Cat. 1	p_{11}	p_{12}	p_{1+}
	Cat. 2	p_{21}	p_{22}	p_{2+}
	Total	p_{+1}	p_{+2}	1.0

Using these frequencies, we can define:

1. An overall proportion of *concordance*:

$$C = \frac{n_{11} + n_{22}}{n}$$

2. Category-specific proportions of concordance:

$$C_1 = \frac{2n_{11}}{2n_{11} + n_{12} + n_{21}}$$

$$C_2 = \frac{2n_{22}}{2n_{22} + n_{12} + n_{21}}.$$

The distinction between concordance and association is that for two responses to be perfectly associated we only require that we can predict the category on one response from the category of the other response, while for two responses to have a perfect concordance, they must fall into the identical category. However, the proportions of concordance, overall or category specific, do not measure agreement.

Among other reasons, they are affected by the marginal totals. Cohen (1960) and others suggest comparing the overall concordance,

$$\theta_1 = \sum_i p_{ii},$$

where p's are the proportions in the above second 2×2 table, with the *chance concordance*,

$$\theta_2 = \sum_i p_{i+}p_{+i},$$

that occurs if the row variable is independent of the column variable, because if two events are independent, the probability of their joint occurrence is the product of their individual or marginal probabilities. This leads to a measure of agreement,

$$\kappa = \frac{\theta_1 - \theta_2}{1 - \theta_2},$$

called the *kappa statistic*, $0 \leq \kappa \leq 1$.

For the case of two categories, κ and its standard error are given by

$$\kappa = \frac{2[n_{11}n_{22} - n_{12}n_{21}]}{n_{1+}n_{+2} + n_{+1}n_{2+}}$$

$$\mathrm{SE}(\kappa) = \frac{2\sqrt{[n_{1+}n_{+1}n_{2+}n_{+2}]/2}}{n^2 - [n_{1+}n_{+1} + n_{2+}n_{+2}]},$$

and the following are guidelines for the evaluation of kappa in clinical research:

$$
\begin{aligned}
\kappa > .75: &\quad \textit{excellent} \text{ reproducibility} \\
.40 \leq \kappa \leq .75: &\quad \textit{good} \text{ reproducibility} \\
0 \leq \kappa < .40: &\quad \textit{marginal/poor} \text{ reproducibility}
\end{aligned}
$$

In general, reproducibility is not good, indicating the need for multiple assessment.

Example 5.1. Two nurses perform ear examaminations focusing on the color of the ear drum (tympanic membrane); each independently assigns each of 100 ears to one of two categories: (1) Normal or gray, or (ii) Not normal (white, pink, orange, or red). The data were:

		Nurse 2		
		Normal	Not Normal	Total
Nurse 1	Normal	35	10	45
	Not Normal	20	35	55
	Total	55	45	100

The results,

$$\kappa = 0.406$$

$$SE(\kappa) = 0.098,$$

indicate that the agreement is barely acceptable.

Note: An SAS program would include these instructions:

```
INPUT N11 N12 N21 N22;
N10=N11+N12; N20=N21+N22; N01=N11+N21; N02=N12+N22;
N=N10+N20; M=N10*N01+N20*N02;
KAPPA=2*(N11*N22 - N12*N21)/(N10*N02+N01*N20);
SE=2*SQRT((N10*N01*N20*N02)/N)/(N*N - M);
CARDS; 35 10 20 35;
PROC PRINT; KAPPA SE;
```

(The next version of SAS, PROC FREQ will have an option for Kappa.)

It should also be pointed out that:

1. Kappa statistic, as a measure for agreement, can also be used when there are more than two categories for classification;

$$\kappa = \frac{\sum_i p_{ii} - \sum_i p_{i+} p_{+i}}{1 - \sum_i p_{i+} p_{+i}}.$$

2. We can form category-specific kappa statistics; e.g., with two categories, we have

$$\kappa 1 = \frac{p_{11} - p_{1+}p_{+1}}{1 - p_{1+}p_{+1}}$$

$$\kappa 2 = \frac{p_{22} - p_{2+}p_{+2}}{1 - p_{2+}p_{+2}}.$$

3. Kappa can be used with stratification. For example, if there are m strata (males and females subjects, or even m pairs of observers), let κ_i and $v_i = \widehat{\mathrm{Var}}(\kappa_i)$ be the kappa and its variance for the ith stratum, and $w_i = 1/v_i$. Define the weighted average:

$$\overline{\kappa} = \frac{\sum w_i \kappa_i}{\sum w_i};$$

then we have:

$$\mathrm{SE}(\overline{\kappa}) = \sqrt{\frac{1}{\sum w_i}}.$$

The chi-square statistic,

$$X^2 = \sum w_i (\kappa_i - \overline{\kappa})^2,$$

can be used to test for homogeneity across strata, with $(m - 1)$ degrees of freedom.

4. The major problem with kappa is that it approaches zero (even with high degree of agreement) if the prevalence goes to 0 or 1.

5.2. PAIR-MATCHED CASE-CONTROL STUDIES

One-to-one matching is a cost-effective design and is perhaps the most popular form used in practice. It is conceptually easy and usually leads to a simple analysis.

5.2.1. The Model

Consider a case-control design and suppose that each individual in a large population has been classified as exposed or not exposed to a certain factor, and as having or not having some disease. The population may then be enumerated in a 2 × 2 table, as follows, with entries being the proportions of the total population:

		Disease		
		+	−	Total
Factor	+	P_1	P_3	$P_1 + P_3$
	−	P_2	P_4	$P_2 + P_4$
	Total	$P_1 + P_2$	$P_3 + P_4$	1

Using these proportions, the association (if any) between the factor and the disease could be measured by the ratio of risks (or relative risk) of being disease positive for those with or without the factor,

$$\begin{aligned}\text{Relative risk} &= \frac{P_1}{P_1 + P_3} \div \frac{P_2}{P_2 + P_4} \\ &= \frac{P_1(P_2 + P_4)}{P_2(P_1 + P_3)},\end{aligned}$$

since in many (although not all) situations, the proportions of subjects classified as disease positive will be small. That is, P_1 is small in comparison with P_3, and P_2 will be small in comparison with P_4. In such a case, the relative risk is almost equal to

$$\begin{aligned}\text{OR} &= \frac{P_1 P_4}{P_2 P_3} \\ &= \frac{P_1/P_3}{P_2/P_4},\end{aligned}$$

the odds ratio of being disease positive, or

$$= \frac{P_1/P_2}{P_3/P_4},$$

the odds ratio of being exposed. This justifies the use of odds ratio to measure differences, if any, in the exposure to a suspected risk factor.

As a technique to control confounding factors in a designed study, individual cases are matched, often one-to-one, to a set of controls chosen to have similar values for the important confounding variables. The most simple example of pair-matched data occurs with a single binary exposure (e.g., smoking versus nonsmoking). The data for outcomes can be represented by a 2×2 table where $(+,-)$ denotes (exposed, unexposed):

		Case		
		+	−	Total
Control	+	n_{11}	n_{01}	$n_{11} + n_{01}$
	−	n_{10}	n_{00}	$n_{10} + n_{00}$
	Total	$n_{11} + n_{10}$	$n_{01} + n_{00}$	n

For example, n_{10} denotes the number of pairs where the case is exposed but the matched control is unexposed. The most suitable statistical model for making inferences about the odds ratio θ is to use the conditional probability of the number of exposed cases among the discordant pairs. Given $(n_{10} + n_{01})$ being fixed, it can be seen that n_{10} has $B(n_{10} + n_{01}, \pi)$ where

$$\pi = \text{OR}/(1 + \text{OR}).$$

5.2.2. The Analysis

Using the above binomial model, with the likelihood function

$$\left\{\frac{\text{OR}}{1 + \text{OR}}\right\}^{n_{10}} \left\{\frac{1}{1 + \text{OR}}\right\}^{n_{01}},$$

it is straightforward to estimate the odds ratio. The results are:

$$\widehat{\text{OR}} = \frac{n_{10}}{n_{01}}$$

$$\widehat{\text{Var}}(\widehat{\text{OR}}) = \frac{n_{10}(n_{10} + n_{01})}{n_{01}^3}.$$

For example, with large samples, a 95 percent confidence interval for the odds ratio is given by

$$\widehat{\text{OR}} \pm (1.96)\{\widehat{\text{Var}}(\widehat{\text{OR}})\}^{1/2}.$$

The null hypothesis of no risk effect can be tested using the score procedure where the z statistic,

$$z = \frac{(n_{10} - n_{01})}{\sqrt{n_{10} + n_{01}}},$$

is compared to percentiles of the standard normal distribution. The corresponding two-tailed procedure based on

$$X^2 = \frac{(n_{10} - n_{01})^2}{n_{10} + n_{01}}$$

is often called the McNemar chi-square test (1 df). It is interesting to note that if we treat a matched pair as a group, or a level of a confounder, and present the data in the form of a 2×2 table,

	Disease Classification		
Exposure	Cases	Controls	Total
Yes	a_i	b_i	$a_i + b_i$
No	c_i	d_i	$c_i + d_i$
Total	1	1	2

then the Mantel–Haenszel method of section 2.4 would yield the same estimate for the odds ratio:

$$\begin{aligned}\widehat{\text{OR}}_{\text{MH}} &= \widehat{\text{OR}} \\ &= \frac{n_{10}}{n_{01}}.\end{aligned}$$

An alternative large-sample method is based on

$$\widehat{\text{Var}}(\log(\widehat{\text{OR}})) = \frac{1}{n_{10}} + \frac{1}{n_{01}},$$

leading to a 95 percent confidence interval of:

$$\frac{n_{10}}{n_{01}} \exp\left[\pm(1.96)\sqrt{\frac{1}{n_{10}} + \frac{1}{n_{01}}}\right].$$

The etiologic fraction or population attributable risk λ is a measure of the impact of an exposure on the population (whereas the relative risk measures the impact of the exposure on the exposed subpopulation). The etiologic fraction is defined as the proportion of disease cases attributable to the risk factor and is expressible as

$$\lambda = p_{1e}\left(1 - \frac{1}{\text{RR}}\right),$$

where p_{1e} is the exposure rate of the subpopulation of cases and RR the relative risk. If the disease is considered as rare (most diseases are rare!), the relative risk can be approximated by the odds ratio, which is estimated by:

$$\widehat{\text{OR}} = \frac{n_{10}}{n_{01}}.$$

In addition, p_{1e} can be estimated using the sample of cases:

$$\widehat{p_{1e}} = \frac{(n_{11} + n_{10})}{n}.$$

Putting these results together, we have

$$\hat{\lambda} = \frac{(n_{11} + n_{10})(n_{10} - n_{01})}{nn_{10}}$$

with:

$$\mathrm{SE}(\hat{\lambda}) = \frac{1}{nn_{10}}\sqrt{\begin{array}{l} n_{11}(n_{10}-n_{01})^2 + \dfrac{(n_{10}^2+n_{11}n_{01})^2}{n_{10}} \\ \quad +n_{01}(n_{11}+n_{10})^2 - \dfrac{(n_{11}+n_{10})^2(n_{10}-n_{01})^2}{n} \end{array}}.$$

For example, with large samples, a 95 percent confidence interval for the attributable risk or etiologic fraction is given by:

$$\hat{\lambda} \pm (1.96)\mathrm{SE}(\hat{\lambda}).$$

Example 5.2. Breslow and Day (1980) used the data on endometrial cancer which were taken from Mack et al. (1976). The investigators identified 63 cases of endometrial cancer occurring in a retirement community near Los Angeles, California from 1971 to 1975. Each diseased individual was matched with $R = 4$ controls who were alive and living in the community at the time the case was diagnosed, who were born within one year of the case, who were of the same marital status and who had entered the community at approximately the same time. The risk factor was previous use of estrogen (yes/no) and the data in the following table were obtained from the first-found matched control; the complete data set with four matched controls will be given in section 5.3:

		Case		
		+	−	Total
Control	+	27	3	30
	−	29	4	33
	Total	66	7	73

An application of the above methods yields

$$\widehat{\mathrm{OR}} = \frac{29}{3}$$
$$= 9.67,$$

and a 95 percent confidence interval for OR is (2.95, 31.74). Similarly, we have

$$\hat{\lambda} = \frac{(27+29)(29-3)}{(63)(29)}$$

$$= .797,$$

and a 95 percent confidence interval for λ is (.639, .954).

5.2.3. The Case of Small Samples

Methods in the previous section are likelihood-based asymptotic methods; they work well when sample sizes are large. However, one of the major reasons that investigators prefer matched design is that they do not have many disease cases. For small studies ($n_{10} + n_{01} \leq 25$, say), a 95 percent confidence interval for OR is obtained as follows:

(i) First, exact 95 percent confidence interval for a binomial parameter π is calculated from the tail probabilities of the binomial distribution using the formulas for lower and upper limits,

$$\pi_L = \frac{n_{10}}{n_{10} + (n_{01} + 1)F_{.975}(2n_{01} + 2, 2n_{10})}$$

$$\pi_U = \frac{(n_{10} + 1)F_{.975}(2n_{10} + 2, 2n_{01})}{n_{01} + (n_{10} + 1)F_{.975}(2n_{10} + 2, 2n_{01})},$$

where $F_{.975}(\nu_1, \nu_2)$ is the 97.5 percentile with ν_1 numerator degrees of freedom and ν_2 denominator degrees of freedom. Then,

(ii) once limits for π are found, they are converted into limit for OR by using the transformation:

$$\text{OR} = \frac{\pi}{1-\pi}.$$

In addition, the p-value for testing

$$H_0 : \text{OR} = 1$$

against

$$H_A : \text{OR} > 1$$

is given by the cumulative binomial probability:

$$\sum_{i=n_{10}}^{n_{10}+n_{01}} \binom{n_{10}+n_{01}}{i} \left(\frac{1}{2}\right)^{n_{10}+n_{01}}.$$

Example 5.3. Referring to the data in Example 5.2 but using the method for small samples, we have

$$\begin{aligned}
\pi_L &= \frac{29}{29+(3+1)F_{.975}(8,58)} \\
&= \frac{29}{29+(3+1)(2.42)} \\
&= .750 \\
\pi_U &= \frac{(29+1)F_{.975}(60,6)}{3+(29+1)F_{.975}(60,6)} \\
&= \frac{(29+1)(4.96)}{3+(29+1)(4.96)} \\
&= .980,
\end{aligned}$$

leading to a 95 percent confidence interval for the odds ratio of $(3.0, 49.6)$.

5.2.4. Risk Factors with Multiple Categories and Ordinal Risks

This subsection is presented here only for completeness; its implementation may be hard for readers in applied fields or students at this level because it involves a matrix inversion and computer packaged programs are not readily available. Consider the pair-matched case-control study with a risk factor having k levels ($k \geq 2$). Let

$$\theta_{ij} = \text{odds ratio for level } i \text{ versus level } j;$$

then the θ_{ij}'s are subject to the constraint

$$\begin{aligned}\theta_{ij} &= \theta_{i1}/\theta_{j1} \\ &= \theta_i/\theta_j, \quad \text{say,}\end{aligned}$$

called the *consistency relationship* between odds ratios. If x_{ij} denotes the number of pairs for which the case has been exposed to level i and the control level j of the risk factor, then it can be shown that the log-likelihood function is given by

$$\ln L = \sum_{i>j} \{x_{ij} \ln(\theta_i/\theta_j) - n_{ij} \ln(1 + \theta_i/\theta_j)\},$$

where

$$n_{ij} = x_{ij} + x_{ji}.$$

Consider the null hypothesis

$$\mathcal{H}_0 : \theta_2 = \cdots = \theta_K = 1.$$

It can be shown that the efficient score chi-square test, called Stuart's test for marginal homogeneity in square tables (Stuart, 1955), can be expressed in a matrix form as follows,

$$\chi^2_{\text{ES}} = (0 - E)^T V^{-1} (0 - E); \qquad df = k - 1,$$

where 0 and E denote the $(K-1)$-dimensional vectors of observed and expected values of the x_{i+}'s (only for $1 \leq i \leq k-1$) while V is the corresponding $(K-1) \times (K-1)$-dimensional covariance matrix.

The elements of E and V are given by

$$\begin{aligned} e_{i+} &= E(x_{i+}) \\ &= (x_{i+} + x_{+i})/2 \\ \text{Var}(x_{i+}) &= (x_{i+} + x_{+i})/4 - x_{ii}/2 \end{aligned}$$

and

$$\text{Cov}(x_{i+}, x_{j+}) = -n_{ij}/4 \qquad \text{for} \quad i \neq j.$$

If the K levels are ordered so that we can assign scores $z_1, z_2, \ldots, z_k$ to these levels, then with an addition of a loglinear model,

$$\theta_i = \exp[\beta(z_i - z_1)],$$

the log-likelihood function becomes

$$\ln L = \sum_{i>j} \{\beta x_{ij}(z_i - z_j) - n_{ij} \ln[1 + e^{\beta(z_i - z_j)}]\},$$

and a test for a "linear trend" (i.e., $\mathcal{H}_0 : \beta = 0$) can be shown to be expressed as

$$\chi^2_{\text{ES}} = \frac{\left[\sum_{i<j}(x_{ij} - x_{ji})(z_i - z_j)\right]^2}{\sum_{i<j} n_{ij}(z_i - z_j)^2}; \qquad 1 \text{ df}.$$

Example 5.4. In the study of endometrial cancer of Example 5.2, using the first control and the following four levels of exposure to conjugated estrogen: (1) none, (2) .1–.299, (3) .3–.625, and (4) $>$ .625 mg, the data for the 59 matched pairs are as follows:

		Control Dose			
		(1)	(2)	(3)	(4)
	(1)	6	2	3	1
Case	(2)	9	4	2	1
Dose	(3)	9	2	3	1
	(4)	12	1	2	1
	Total	36	9	10	4

We obtain the test statistic

$$\chi^2_{\mathrm{ES}} = (36-24, 9-12.5, 10-12.5)$$
$$\times \begin{bmatrix} 9 & -2.75 & -3 \\ -2.75 & 4.25 & -1 \\ -3 & -1 & -4.75 \end{bmatrix}^{-1} \begin{bmatrix} 36-24 \\ 9-12.5 \\ 10-12.5 \end{bmatrix}$$
$$= 16.96,$$

which is highly significant by reference to tables of chi-square with three degrees of freedom. Assigning scores $z_1 = 1$, $z_2 = 2$, $z_3 = 3$, $z_4 = 4$ to the four exposure levels, we have a one-degree-of-freedom, highly significant test statistic for the trend

$$\chi^2_{\mathrm{ES}} = 14.71.$$

5.3. MULTIPLE MATCHING

One-to-one matching is a cost-effective design. However, an increase in the number of controls may give the study more power. In epidemiologic studies, there are typically a small number of cases and a large number of potential controls to select from. When the controls are more easily available than cases, it is more efficient and effective to select more controls for each case. Breslow and Day (1980) proved that the efficiency of an M-to-one control-case ratio for estimating a relative risk relative to having complete information on the control population (i.e., $M = \infty$) is $M/(M+1)$. Hence, a one-to-one matching is 50 percent efficient, four-to-one matching is 80 percent efficient, five-to-one matching is 83 percent efficient, and so on. The gain in efficiency is rapidly diminishing for designs with $M \geq 5$.

5.3.1. The Conditional Approach

The analysis of one-to-one matching design was conditional on the number of pairs showing differences in the exposure history, the $(-,+)$ and $(+,-)$ cells. Similarly, considering an M-to-one matching design, we will use a conditional approach, fixing the number m of exposed individuals in a matched set; and the sets with $m = 0$ or $m = M + 1$ (similar to $(-,-)$ and $(+,+)$ cells in the one-to-one matching design) will be ignored.

If we fix the number of exposed individuals in each stratum, then it can be straightforwardly shown that

$$\begin{aligned}&\Pr(\text{case exposed}/m \text{ exposed in a stratum})\\&\quad = \frac{(m)(\text{OR})}{(m)(\text{OR}) + M - m + 1},\end{aligned}$$

where OR is the odds ratio representing the effect of exposure. The result for pair-matched design in section 5.2.1 is a special case where $M = m = 1$.

For the strata, or matched sets, with exactly m $(m = 1,2,\ldots,M)$ exposed individuals, let

$$\begin{aligned}n_{1,m-1} &= \text{number of sets with an exposed case}\\ n_{0,m} &= \text{number of sets with an unexposed case}\\ n_m &= n_{1,m-1} + n_{0,m}.\end{aligned}$$

Then, given n_m being fixed, $n_{1,m-1}$ has $B(n_m, p_m)$ where

$$p_m = \frac{(m)(\text{OR})}{(m)(\text{OR}) + M - m + 1}.$$

5.3.2. Estimation of the Odds Ratio

From the joint (conditional) likelihood function,

$$\begin{aligned}L(\text{OR}) = \prod_{m=1}^{M} &\left(\frac{(m)(\text{OR})}{(m)(\text{OR}) + M - m + 1}\right)^{n_{1,m-1}}\\ &\times \left(\frac{M - m + 1}{(m)(\text{OR}) + M - m + 1}\right)^{n_{0,m}}\end{aligned}$$

or

$$\ln L = \sum_{m=1}^{M} \{n_{1,m-1} \ln[(m)(\text{OR})] + n_{0,m} \ln(M - m + 1)$$
$$- n_m \ln[(m)(\text{OR}) + M - m + 1]\},$$

one can obtain the MLE of OR, but such a solution requires an iterative procedure and a computer algorithm. We will come back to this topic with a complete solution in section 5.4.

A simple method for estimating the odds ratio would be to treat a matched set, consisting of one case and M matched controls, as a stratum (i.e., a level of some confounder). We then present the data from this stratum in the form of a 2×2 table,

	Disease Classification		
Exposure	Cases	Controls	Total
Yes	a	b	$a+b$
No	c	d	$c+d$
Total	1	M	$M+1$

and obtain the Mantel–Haenszel estimate for the odds ratio:

$$\widehat{\text{OR}}_{\text{MH}} = \frac{\sum \frac{ad}{M+1}}{\sum \frac{bc}{M+1}}.$$

The result turns out quite simple:

$$\widehat{\text{OR}}_{\text{MH}} = \frac{\sum (M - m + 1) n_{1,m-1}}{\sum m n_{0,m}}.$$

The Mantel–Haenszel estimate has been widely used in the analysis of case-control studies with multiple matching (see Example 5.5).

5.3.3. Testing for Exposure Effect

From the likelihood function of section 5.3.2,

$$L = \prod_{m=1}^{M} \left(\frac{(m)(\mathrm{OR})}{(m)(\mathrm{OR}) + M - m + 1} \right)^{n_{1,m-1}} \left(\frac{M - m + 1}{(m)(\mathrm{OR}) + M - m + 1} \right)^{n_{0,m}},$$

it can be verified that the score test for

$$\mathcal{H}_0 : \theta = 1$$

is given by

$$\chi^2_{\mathrm{ES}} = \frac{\left[\sum_{m=1}^{M} \left(n_{1,m-1} - \dfrac{mn_m}{M+1} \right) \right]^2}{\dfrac{1}{(M+1)^2} \sum_{m=1}^{M} mn_m (M - m + 1)},$$

a chi-square test with one degree of freedom.

Example 5.5. Breslow and Day (1980) used the data on endometrial cancer which were taken from Mack et al. (1976). The investigators identified 63 cases of endometrial cancer occurring in a retirement community near Los Angeles, California from 1971 to 1975. Each diseased individual was matched with $R = 4$ controls who were alive and living in the community at the time the case was diagnosed, who were born within one year of the case, who were of the same marital status and who had entered the community at approximately the same time. The risk factor was previous use of estrogen (yes/no) and the data in Example 5.2 involve only the first-found matched control. We are now able to analyze the complete data set (*4-to-1 matching*):

	Number of Exposed Individuals in Each Matched Set			
Case	1	2	3	4
Exposed	4	17	11	9
Unexposed	6	3	1	1
Total	10	20	12	10

Using these 52 sets with four matched controls, we have:

$$\chi^2_{\text{ES}} = \frac{(25)\left[\left(4-\frac{1\times 10}{5}\right)+\left(17-\frac{2\times 20}{5}\right)+\left(11-\frac{3\times 12}{5}\right)+\left(9-\frac{4\times 10}{5}\right)\right]^2}{(1\times 10\times 4)+(2\times 20\times 3)+(3\times 12\times 2)+(4\times 10\times 1)}$$

$$= 22.95.$$

The Mantel–Haenszel estimate for the odds ratio is:

$$\widehat{\text{OR}}_{\text{MH}} = \frac{(4)(4)+(3)(17)+(2)(11)+(1)(9)}{(1)(6)+(2)(3)+(3)(1)+(4)(1)}$$
$$= 5.16.$$

When the number of controls matched to a case, M, is variable (mostly due to missing data), the test for exposure effects should incorporate data from all strata:

$$\chi^2_{\text{ES}} = \frac{\left[\sum_M \sum_{m=1}^{M} \left(n_{1,m-1} - \frac{mn_m}{M+1}\right)\right]^2}{\sum_M \sum_{m=1}^{M} \frac{mn_m(M-m+1)}{(M+1)^2}}.$$

The corresponding Mantel–Haenszel estimate for the odds ratio is:

$$\widehat{\text{OR}}_{\text{MH}} = \frac{\sum_M \sum[(M-m+1)n_{1,m-1}]/(M+1)}{\sum_M \sum mn_{0,m}/(M+1)}.$$

Example 5.6. Refer to the data on endometrial cancer of Example 5.5; due to missing data, we have some cases matching to four and some matching to three controls. In addition to *4-to-1 matching,*

	Number of Exposed Individuals in Each Matched Set			
Case	1	2	3	4
Exposed	4	17	11	9
Unexposed	6	3	1	1
Total	10	20	12	10

we also have *3-to-1 matching*:

Case	Number of Exposed Individuals in Each Matched Set		
	1	2	3
Exposed	1	3	0
Unexposed	0	0	0
Total	1	3	0

With the inclusion of the four sets having three matched controls, the result of Example 5.5 becomes:

$$\chi^2_{ES} = \frac{\left[\left\{\left(4-\frac{1\times 10}{2}\right)+\left(17-\frac{2\times 20}{5}\right)+\left(11-\frac{3\times 12}{5}\right)+\left(9-\frac{4\times 10}{5}\right)\right\}+\left\{\left(1-\frac{1\times 1}{4}\right)+\left(3-\frac{2\times 3}{4}\right)\right\}\right]^2}{\frac{1}{25}[(1\times 10\times 4)+(2\times 20\times 3)+(3\times 12\times 2)+(4\times 10\times 1)]+\frac{1}{16}[1\times 1\times 3+2\times 3\times 2]}$$

$$= 27.57.$$

The Mantel–Haenszel estimate for the odds ratio is:

$$\widehat{OR}_{MH} = \frac{[(4)(4)+(3)(17)+(2)(11)+(1)(9)]/(5)+[(3)(1)+(2)(3)+(1)(0)]/(4)}{[(1)(6)+(2)(3)+(3)(1)+(4)(1)]/(5)+[(1)(0)+(2)(0)+(3)(0)]/(4)}$$

$$= 5.75.$$

5.3.4. Testing for Homogeneity

This subsection is concerned with an important aspect of case-control studies, the homogeneity of the odds ratio. To estimate the odds ratio (as seen in subsection 5.3.1), we made the assumption that the exposure probabilities of a risk factor are allowed to differ from matched set to matched set (due to confounding effects) but the odds ratio is constant across matched sets. This homogeneity can be tested using the statistic

$$S = \sum_{m=1}^{M} \hat{p}_m(n_m \hat{p}_m - n_{1,m-1}),$$

where $\hat{p}_m$, $1 \leq m \leq M$, is the probability that the case has been exposed given that there are m exposed members of a set,

$$\hat{p}_m = \frac{m\widehat{\text{OR}}}{m\widehat{\text{OR}} + M - m + 1}.$$

The estimated variance of S is given by:

$$\widehat{\text{Var}}(S) = \sum_{m=1}^{M} n_m \hat{p}_m^3 \hat{q}_m - \frac{[\sum_{m=1}^{M} n_m \hat{p}_m^2 \hat{q}_m]^2}{\sum_{m=1}^{M} n_m \hat{p}_m \hat{q}_m}.$$

The standardized statistic $\hat{S}/\text{Var}(\hat{S})^{1/2}$ is distributed asymptotically as standard normal (Liang and Self, 1985).

Example 5.7. Referring to the 4-to-1 matched data of Example 5.5 and the Mantel–Haenszel estimate for the odds ratio ($\text{OR}_{\text{MH}} = 5.16$), we have:

$$S = -0.056$$

$$\widehat{\text{Var}}(S) = 4.23 - \frac{(5.58)^2}{7.60}$$

$$= 0.133,$$

leading to:

$$z = \frac{S}{\{\text{Var}(S)\}^2}$$

$$= 0.15,$$

indicating that the homogeneity of the odds ratio fits rather well.

5.4. CONDITIONAL LOGISTIC REGRESSION

Recall from Chapter 4 that, in a variety of applications using regression analysis, the dependent variable of interest has only two possible outcomes, and therefore can be represented by an indicator variable taking on values 0 and 1. An important application is the analysis of case-control studies where the dependent variable represents the disease status, 1 for a case and 0 for a control. The methods that have been widely and successfully used for these applications are based on the logistic model. This section also deals with the cases where the dependent variable of interest is binary following a binomial distrubution—the same as those using logistic regression analyses, but data are matched. Again, the term *matching* refers to the pairing of one or more controls to each case on the basis of their similarity with respect to selected variables used as *matching criteria*, as seen in sections 5.2 and 5.3. Although the primary objective of matching is the elimination of biased comparison between cases and controls, this objective can be accomplished only if matching is followed by an analysis that corresponds to the matched design. Unless the analysis properly accounts for the matching used in the selection phase of a case-control study, the results can be biased. In other words, matching (which refers to the selection process) is only the first step of a two-step process that can be used effectively to control for confounders: (1) matching design, followed by (2) matched analysis. Suppose the purpose of the research is to assess relationships between the disease and a set of covariates using a matched case-control design; the regression techniques for the statistical analysis of such relationships is based on the *conditional logistic model.*

The following are two typical examples; the first one is a case-control study of vaginal carcinoma which involves two binary risk factors. Of course, we can investigate one covariate at a time using the method of section 5.3.

Example 5.8. Consider the data taken from a study by Herbst et al. (1971); the cases were eight women 15 to 22 years of age who were diagnosed with vaginal carcinoma between 1966 and 1969. For each case, four controls were found in the birth records of patients having their babies delivered within five days of the case in the same hospital.

The risk factors of interest are the mother's bleeding in this pregnancy (N = No, Y = Yes) and any previous pregnancy loss by the

TABLE 5.1. Vaginal Carcinoma Data

		Responses (Bleeding, Previous Loss)			
		Control Subject Number			
Set	Case	1	2	3	4
1	(N,Y)	(N,Y)	(N,N)	(N,N)	(N,N)
2	(N,Y)	(N,Y)	(N,N)	(N,N)	(N,N)
3	(Y,N)	(N,Y)	(N,N)	(N,N)	(N,N)
4	(Y,Y)	(N,N)	(N,N)	(N,N)	(N,N)
5	(N,N)	(Y,Y)	(N,N)	(N,N)	(N,N)
6	(Y,Y)	(N,N)	(N,N)	(N,N)	(N,N)
7	(N,Y)	(N,Y)	(N,N)	(N,N)	(N,N)
8	(N,Y)	(N,N)	(N,N)	(N,N)	(N,N)

mother (N = No, Y = Yes). The data are given in Table 5.1, as used by Holford (1982).

In the second example, one of the four covariates is on a continuous scale.

Example 5.9. The data were taken from Hosmer and Lemeshow (1989) in a study of low-birthweight babies (the cases); only a small portion of the data set is presented here for illustration (Table 5.2). For each of the first 15 cases, we retain only the first three matched controls (even though the number of controls per case need not be the same). Four risk factors are under investigation: weight (in pounds) of the mother at the last menstrual period, hypertension, smoking, and uterine irritability; for the last three factors, a value of 1 indicates a yes and a value of 0 indicates a no. The mother's age was used as the matching variable.

5.4.1. Simple Regression Analysis

In this section we will discuss the basic ideas of simple regression analysis when only one predictor or independent variable is available for predicting the binary response of interest. We illustrate these for the more simple designs in which each matched set has one case and case i is matched to m_i controls; the number of controls m_i may vary from case to case.

TABLE 5.2. Low Birthweight Data

Matched Set	Case	MotherWeight	Hypertension	Smoking	U-Irritability
1	1	130	0	0	0
	0	112	0	0	0
	0	135	1	0	0
	0	270	0	0	0
2	1	110	0	0	0
	0	103	0	0	0
	0	113	0	0	0
	0	142	0	1	0
3	1	110	1	0	0
	0	100	1	0	0
	0	120	1	0	0
	0	229	0	0	0
4	1	102	0	0	0
	0	182	0	0	1
	0	150	0	0	0
	0	189	0	0	0
5	1	125	0	0	1
	0	120	0	0	1
	0	169	0	0	1
	0	158	0	0	0
6	1	200	0	0	1
	0	108	1	0	1
	0	185	1	0	0
	0	110	1	0	1
7	1	130	1	0	0
	0	95	0	1	0
	0	120	0	1	0
	0	169	0	0	0
8	1	97	0	0	1
	0	128	0	0	0
	0	115	1	0	0
	0	190	0	0	0
9	1	132	0	1	0
	0	90	1	0	0
	0	110	0	0	0
	0	133	0	0	0
10	1	105	0	1	0
	0	118	1	0	0
	0	155	0	0	0
	0	241	0	1	0

TABLE 5.2. *(Continued)*

Matched Set	Case	MotherWeight	Hypertension	Smoking	U-Irritability
11	1	96	0	0	0
	0	168	1	0	0
	0	160	0	0	0
	0	133	1	0	0
12	1	120	1	0	1
	0	120	1	0	0
	0	167	0	0	0
	0	250	1	0	0
13	1	130	0	0	1
	0	150	0	0	0
	0	135	0	0	0
	0	154	0	0	0
14	1	142	1	0	0
	0	153	0	0	0
	0	110	0	0	0
	0	112	0	0	0
15	1	102	1	0	0
	0	215	1	0	0
	0	120	0	0	0
	0	150	1	0	0

The Likelihood Function

In our framework, let x_i be the covariate value for case i and x_{ij} be the covariate value for the jth control matched to case i. Then, for the ith matched set, the conditional probability of the observed outcome (that the subject with covariate value x_i be the case) given that we have one case per matched set is

$$\frac{\exp(\beta x_i)}{\exp(\beta x_i) + \sum_j^{m_i} \exp(\beta x_{ij})}.$$

If the sample consists of N matched sets, then the conditional likelihood function is the product of the above terms over the N matched sets,

$$L = \prod_{i=1}^{N} \frac{\exp(\beta x_i)}{\exp(\beta x_i) + \sum_j^{m_i} \exp(\beta x_{ij})},$$

from which we can obtain maximum likelihood estimate of the parameter β (Breslow, 1982).

Measure of Association

Similar to the case of the logistic model, $\exp(\beta)$ represents

(i) the Odds Ratio associated with an exposure if X is binary (exposed $X = 1$ vs. unexposed $X = 0$), or

(ii) the Odds Ratio due to one unit increase if X is continuous ($X = x + 1$ vs. $X = x$),

and after $\hat{\beta}$ and its standard error have been obtained, a 95 percent confidence interval for the above odds ratio is given by:

$$\exp[\hat{\beta} \pm 1.96\mathrm{SE}(\hat{\beta})].$$

A Special Case

Consider now the most simple case of a pair-matched (i.e., one-to-one matching) with a binary covariate: exposed $X = 1$ versus unexposed $X = 0$. Let the data be summarized and presented as in section 5.2.

Control	Case 1	Case 0
1	n_{11}	n_{01}
0	n_{10}	n_{00}

For example, n_{10} denotes the number of pairs where the case is exposed but the matched control is unexposed. The above likelihood function is reduced to

$$L(\beta)$$

$$= \left\{\frac{1}{1+1}\right\}^{n_{00}} \cdot \left\{\frac{\exp(\beta)}{1+\exp(\beta)}\right\}^{n_{10}} \left\{\frac{1}{1+\exp(\beta)}\right\}^{n_{01}} \cdot \left\{\frac{\exp(\beta)}{\exp(\beta)+\exp(\beta)}\right\}^{n_{11}}$$

$$= \frac{\exp(\beta n_{10})}{(1+\exp(\beta))^{n_{10}+n_{01}}}.$$

From this we can obtain a point estimate,

$$\hat{\beta} = \frac{n_{10}}{n_{01}},$$

which is the usual odds ratio estimate from pair-matched data identical to that of subsection 5.2.2.

Tests of Association

Another aspect of statistical inference concerns the test of significance; the null hypothesis to be considered is:

$$\mathcal{H}_0 : \beta = 0.$$

The reason for interest in testing whether or not $\beta = 0$ is that $\beta = 0$ implies there is no relation between the binary dependent variable and the covariate X under investigation. Since the likelihood function is rather simple, one can easily derive, say, the score test for the above null hypothesis; however, nothing would be gained by going through this exercise. We can simply apply a McNemar chi-square test (if the covariate is binary or categorical; see subsections 5.2.2 and 5.2.4) or a paired t-test or signed-rank Wilcoxon test (if the covariate under investigation is on a continuous scale). Of course, the application of the conditional logistic model is still desirable, at least in the case of a continuous covariate, because it would provide a measure of association—the odds ratio.

Example 5.10. Refer to the data for low-birthweight babies in Example 5.9 and suppose we want to investigate the relationship between the low-birthweight problem, our outcome for the study, and the weight of the mother taken at the last menstrual period. An application of the simple conditional logistic regression analysis yields:

Variable	Coefficient	St. Error	z Statistic	p-Value
MotherWeight	−0.0211	0.0112	−1.884	0.0593

The result indicates that the effect of the mother's weight is nearly significant at the 5 percent level ($p = 0.0593$).

The odds ratio associated with, say ten pounds (10 lbs) increase in weight is

$$\exp(-.2114) = 0.809.$$

If a mother increases her weight about ten pounds, the odds to have a low-birthweight baby are reduced by almost 20 percent.

Note: An SAS program would include these instructions:

```
INPUT SET CASE MWEIGHT;
DUMMYTIME=2-CASE;
CARDS;
(data)
PROC PHREG DATA=LOWWEIGHT;
MODEL=DUMMYTIME*CASE(0)=MWEIGHT/TIES=DISCRETE;
STRATA=SET;
```

where LOWWEIGHT is the name assigned to the data set, DUMMYTIME is the name for the makeup time variable defined in the upper part of the program, CASE is the case-control status indicator (coded as 1 for a case and 0 for a control), and MWEIGHT is the variable name for weight of the mother at the last menstrual period. The matched SET number (1 to 15 in this example) is used as the stratification factor.

5.4.2. Multiple Regression Analysis

The effect of some factor on a dependent or response variable may be influenced by the presence of other factors through effect modifications (i.e., interactions). Therefore, in order to provide a more comprehensive analysis, it is very desirable to consider a large number of factors and sort out which ones are most closely related to the dependent variable. In this section, we will discuss a multivariate method for such a risk determination. This method, which is multiple conditional logistic regression analysis, involves a linear combination of the explanatory or independent variables; the variables must be quantitative with particular numerical values for each patient. A covariate or independent variable—such as a patient characteristic—may be dichotomous, polytomous, or continuous (categorical factors will be represented by dummy variables). Examples of dichotomous covariates are sex, and presence or absence of certain co-morbidity. Polytomous covariates include race, and different grades of symptoms; these can be covered by the use of dummy variables. Continuous covariates include patient age, blood pressure, and so on. In many cases, data transformations (e.g., taking the logarithm) may be desirable to satisfy the linearity assumption. We illustrate this process for a very general design in which matched set i, $(1 \leq N)$, has n_i cases and cases which were matched to m_i controls; the numbers of cases n_i and controls m_i vary from matched set to matched set.

The Likelihood Function

For the general case of n_i cases matched to m_i controls in a set, we have the conditional probability of the observed outcome (that a specific set of n_i subjects are cases) given that the number of cases is n_i (any n_i subjects could be cases),

$$\frac{\exp\left(\sum_{j=1}^{n_i}[\beta^T \mathbf{x_j}]\right)}{\sum_{R(n_i,m_i)} \exp\left(\sum_{j=1}^{n_i}[\beta^T \mathbf{x_j}]\right)},$$

where the sum in the denominator ranges over the collections $R(n_i, m_i)$ of all partitions of the $(n_i + m_i)$ subjects into two, one of size n_i and one of size m_i. The full conditional likelihood is the product over all matched sets, one probability for each set, i.e.,

$$L = \prod_{i=1}^{N} \frac{\exp\left(\sum_{j=1}^{n_i}[\beta^T \mathbf{x_j}]\right)}{\sum_{R(n_i,m_i)} \exp\left(\sum_{j=1}^{n_i}[\beta^T \mathbf{x_j}]\right)}$$

(Breslow, 1982).

From the above result, it follows that the full likelihood of the observed outcome has the same mathematical form as the overall partial likelihood for the proportional hazards survival model with strata, one for each matched set, and one event time for each (more in Chapter 7). This enables us to adapt programs written for the proportional hazards model to analyze epidemiologic matched studies as seen in subsequent examples. The essential features of the adaptation are:

(i) creating matched set numbers and using them as different levels of a stratification factor, and
(ii) assigning a number to each subject; these numbers will be used in the place of duration times. These numbers are chosen arbitrarily as long as the number assigned to a case is smaller than the number assigned to a control in the same matched set. This is possible because when there is only one event in each set, the numerical value for the time to event becomes irrelevant.

Similar to the univariate case, $\exp(\beta_i)$ represents:

(i) the Odds Ratio associated with an exposure if X_i is binary (exposed $X_i = 1$ vs. unexposed $X_i = 0$), or
(ii) the Odds Ratio due to one unit increase if X_i is continuous ($X_i = x + 1$ vs. $X_i = x$),

and after $\hat{\beta}_i$ and its standard error have been obtained, a 95 percent confidence interval for the above odds ratio is given by:

$$\exp[\hat{\beta}_i \pm 1.96\mathrm{SE}(\hat{\beta}_i)].$$

These results are necessary in the effort to identify important risk factors in matched designs. Of course, before such analyses are done, the problem and the data have to be examined carefully. If some of the variables are highly correlated, then one or fewer of the correlated factors are likely to be as good predictors as all of them; information from other similar studies also has to be incorporated so as to drop some of these correlated explanatory variables. The uses of products, such as X_1X_2, and higher-power terms, such as X_1^2, may be necessary and can improve the goodness-of-fit (unfortunately, it is very hard to tell!). It is important to note that we are assuming a *linear* regression model in which, for example, the odds ratio due to one unit increase in the value of a continuous X_i ($X_i = x + 1$ vs. $X_i = x$) is independent of x. Therefore, if this *linearity* seems to be violated (again, it is very hard to tell; the only easy way is fitting a polynomial model as seen in a later example), the incorporation of powers of X_i should be seriously considered. The use of products will help in the investigation of possible effect modifications. And, finally, there is the messy problem of missing data; most packaged programs would delete a subject if one or more covariate values are missing.

Testing Hypotheses in Multiple Regression

Once we have fit a multiple conditional logistic regression model and obtained estimates for the various parameters of interest, we want to answer questions about the contributions of various factors to the prediction of the binary response variable using matched designs. There are three types of such questions:

(i) An overall test: Taken collectively, does the entire set of explanatory or independent variables contribute significantly to the prediction of response?

(ii) Test for the value of a single factor: Does the addition of one particular variable of interest add significantly to the prediction of response over and above that achieved by other independent variables?

(iii) Test for contribution of a group of variables: Does the addition of a group of variables add significantly to the prediction of response over and above that achieved by other independent variables?

Overall Regression Test We now consider the first question stated above concerning an overall test for a model containing k factors. The null hypothesis for this test may stated as: "all k independent variables *considered together* do not explain the variation in the response any more than the size alone." In other words,

$$\mathcal{H}_0 : \beta_1 = \beta_2 = \cdots = \beta_k = 0.$$

Three statistics can be used to test this *global* null hypothesis; each has an symptotic chi-squared distribution with k degrees of freedom under $\mathcal{H}_0$.

(i) Likelihood ratio test:

$$\chi^2_{\mathrm{LR}} = 2[\ln L(\hat{\beta}) - \ln L(\mathbf{0})].$$

(ii) Wald's test:

$$\chi_W^2 = \hat{\beta}^T [\hat{V}(\hat{\beta})]^{-1} \hat{\beta}.$$

(iii) Score test:

$$\chi_S^2 = \left[\frac{\delta \ln L(\mathbf{0})}{\delta \beta}\right] \left[-\frac{\delta^2 \ln L(\mathbf{0})}{\delta \beta^2}\right]^{-1} \left[\frac{\delta \ln L(\mathbf{0})}{\delta \beta}\right].$$

All three statistics are provided by most standard computer programs such as SAS and they are asymptotically equivalent (i.e., for very large sample sizes), yielding identical statistical decisions most of the time; however, Wald's test is much less often used than the other two. □

Example 5.11. Referring to the data for low-birthweight babies in Example 5.9 with all four covariates, we have the following test statistics for the global null hypothesis:

(i) Likelihood test:

$$\chi_{LR}^2 = 9.530 \quad \text{with} \quad 4 \text{ dfs}; \qquad p = 0.0491.$$

(ii) Wald's test:

$$\chi_W^2 = 6.001 \quad \text{with} \quad 4 \text{ dfs}; \qquad p = 0.1991.$$

(iii) Score test:

$$\chi_S^2 = 8.491 \quad \text{with} \quad 4 \text{ dfs}; \qquad p = 0.0752.$$

The results indicate a weak combined explanatory power; Wald's test is not even significant. Very often, this implicitly means that maybe only one or two covariates are significantly associated to the response of interest (a weak overall correlation).

Test for a Single Variable Let us assume that we now wish to test whether the addition of one particular independent variable of interest adds significantly to the prediction of the response over and above that achieved by other factors already present in the model (usually after seeing a significant result for the above global hypothesis). The null hypothesis for this single-variable test may stated as: "Factor X_i does not have any value added to the prediction of the response *given that other factors are already included in the model.*" In other words,

$$\mathcal{H}_0 : \beta_i = 0.$$

To test such a null hypothesis, one can use

$$z_i = \frac{\hat{\beta}_i}{\mathrm{SE}(\hat{\beta}_i)},$$

where $\hat{\beta}_i$ is the corresponding estimated regression coefficient and $\mathrm{SE}(\hat{\beta}_i)$ is the estimate of the standard error of $\hat{\beta}_i$, both of which are printed by standard computer packaged programs such as SAS. In performing this test, we refer the value of the z score to percentiles of the standard normal distribution; for example, we compare the absolute value of z to 1.96 for a two-sided test at the 5 percent level. □

Example 5.12. Referring to the data for low-birthweight babies in Example 5.9 with all four covariates, we have the following results. Only the mother's weight ($p = 0.0942$) and urine irritability ($p = 0.0745$) are marginally significant. In fact, these two variables are highly correlated, so that if one is deleted from the model, the other would become more significant.

Variable	Coefficient	St. Error	z Statistic	p-Value
MotherWeight	−0.0191	0.0114	−1.673	0.0942
Smoking	−0.0885	0.8618	−0.103	0.9182
Hypertension	0.6325	1.1979	0.528	0.5975
Urine Irritability	2.1376	1.1985	1.784	0.0745

The overall tests and the tests for single variables are implemented simultaneously using the same computer program, and here is another example:

Example 5.13. Referring to the data for vaginal carcinoma in Example 5.8, an application of a conditional logistic regression analysis yields the following results:

(i) Likelihood test for the global hypothesis:

$$\chi^2_{\text{LR}} = 9.624 \quad \text{with} \quad 2 \text{ dfs}; \quad p = 0.0081.$$

(ii) Wald's test for the global hypothesis:

$$\chi^2_W = 6.336 \quad \text{with} \quad 2 \text{ dfs}; \quad p = 0.0027.$$

(iii) Score test for the global hypothesis:

$$\chi^2_S = 11.860 \quad \text{with} \quad 2 \text{ dfs}; \quad p = 0.0421.$$

And for individual covariates:

Variable	Coefficient	St. Error	z Statistic	p-Value
Bleeding	1.6198	1.3689	1.183	0.2367
Pregnancy Loss	1.7319	0.8934	1.938	0.0526

In addition to a priori interest in the effects of individual covariates, given a continuous variable of interest, one can fit a polynomial model and use this type of test to check for linearity. It can also be used to check for a single product representing an effect modification.

Example 5.14. Refer to the data for low-birthweight babies in Example 5.9, but this time we investigate only one covariate, the mother's weight. After fitting the second-degree polinomial model, we obtained a result which indicates that the *curvature effect* is negligible ($p = 0.9131$).

Contribution of a Group of Variables This testing procedure addresses the more general problem of assessing the additional contribution of two or more factors to the prediction of the response over and above that made by other variables already in the regression model. In other words, the null hypothesis is of the form

$$\mathcal{H}_0 : \beta_1 = \beta_2 = \cdots = \beta_m = 0.$$

To test such a null hypothesis, one can perform a likelihood ratio chi-squared test, with m dfs,

$$\chi^2_{\text{LR}} = 2[\ln L(\hat{\beta};\text{ all } X\text{'s}) - \ln L(\hat{\beta};\text{ all other } X\text{'s with } X\text{'s under investigation deleted})]. \qquad \square$$

As with the above tests for individual covariates, this *multiple contribution* procedure is very useful for assessing the importance of potential explanatory variables. In particular, it is often used to test whether a similar group of variables, such as *demographic characteristics*, is important for the prediction of the response; these variables have some trait in common. Another application would be a collection of powers *and/or* product terms (referred to as interaction variables). It is often of interest to assess the interaction effects collectively before trying to consider individual interaction terms in a model as previously suggested. In fact, such use reduces the total number of tests to be performed and this, in turn, helps to provide better control of overall Type I error rates which may be inflated due to multiple testing.

Example 5.15. Refer to the data for low-birthweight babies in Example 5.9 with all four covariates. We consider, collectively, these three interaction terms: MotherWeight∗Smoking, MotherWeight∗Hypertension, MotherWeight∗Urine Irritability. The basic idea is to see if *any* of the other variables would modify the effect of the mother's weight on the response (having a low-birthweight baby).

1. With the original four variables, we obtained: $\ln L = -16.030$.
2. With all seven variables, four original plus three products, we obtained: $\ln L = -14.199$.

Therefore we have

$$\begin{aligned}\chi^2_{\text{LR}} &= 2[\ln L(\hat{\beta};\ \text{seven variables}) - \ln L(\hat{\beta};\ \text{four original variables})]\\ &= 3.662; \quad 2\ \text{dfs}, \quad p\text{-value} \geq .10,\end{aligned}$$

indicating a rather weak level of interactions.

Stepwise Regression In many applications, our major interest is to identify important risk factors. In other words, we wish to identify from many available factors a small subset of factors that relate significantly to the outcome (e.g., the disease under investigation). In that identification process, of course, we wish to avoid a large Type I (false positive) error. In a regression analysis, a Type I error corresponds to including a predictor that has no real relationship to the outcome; such an inclusion can greatly confuse the interpretation of the regression results. In a standard multiple regression analysis, this goal can be achieved by using a strategy that adds into or removes from a regression model one factor at a time according to a certain order of relative importance. Therefore, the two important steps are:

1. Specifying a criterion or criteria for selecting a model.
2. Specifying a strategy for applying the chosen criterion or criteria.

The process follows the same outline of Chapter 4 for logistic regression, that we combine the forward selection and backward elimination in the stepwise process and selection at each step is based on the likelihood ratio chi-square test. The SAS PROC PHREG does have an automatic stepwise option to implement these features. □

Example 5.16. Refer to the data for low-birthweight babies in Example 5.9 with all four covariates: mother's weight, smoking, hypertension, and urine irritability. This time we perform a stepwise regression analysis in which we specify that a variable has to be significant at the 0.10 level before it can enter into the model and that a variable in the model has to be significant at the 0.15 level for it to remain in the model (most standard computer programs allow users to make these selections; default values are available). First, we get these individual score test results for all variables:

Variable	Score χ^2	p-Value
MotherWeight	3.9754	0.0462
Smoking	0.0000	1.0000
Hypertension	0.2857	0.5930
Urine Irritability	5.5556	0.0184

These indicate that Urine Irritability is the most significant variable; thus:

Step 1: Variable Urine Irritability is entered.
Analysis of Variables Not in the Model:

Variable	Score χ^2	p-Value
MotherWeight	2.9401	0.0864
Smoking	0.0027	0.9584
Hypertension	0.2857	0.5930

Step 2: Variable MotherWeight is entered.
Analysis of Variables in the Model:

Factor	Coefficient	St. Error	z Statistic	p-Value
MotherWeight	−0.0192	0.0116	−1.655	0.0978
Urine Irritability	2.1410	1.1983	1.787	0.0740

Neither variable is removed.
Analysis of Variables Not in the Model:

Variable	Score χ^2	p-Value
Smoking	0.0840	0.7720
Hypertension	0.3596	0.5487

No (additional) variables meet the 0.1 level for entry into the model.

Note: An SAS program would include these instructions:

```
PROC PHREG DATA LOWWEIGHT;
MODEL DUMMYTIME*CASE(0)= MWEIGHT SMOKING HYPERT UIR-
RIT/SELECTION=STEPWISE SLENTRY=.10 SLSTAY=.15;
STRATA=SET;
```

(HYPERT and UIRRIT are hypertension and urine irritability.) The default values for SLENTRY (p-value to enter) and SLSTAY (p-value to stay) are .05 and .10, respectively.

5.5. EXERCISES

1. It has been noted that metal workers have an increased risk for cancer of the internal nose and paranasal sinuses, perhaps as a result of exposure to cutting oils. Therefore, a study was conducted to see whether this particular exposure also increases the risk for squamous cell carcinoma of the scrotum (Rousch et al., 1982).

Cases included all 45 squamous cell carcinomas of the scrotum diagnosed in Connecticut residents from 1955 to 1973, as obtained from the Connecticut Tumor Registry. Matched controls were selected for each case based on the age at death (within 8 years), year of death (within 3 years), and number of jobs as obtained from combined death certificate and Directory sources. An occupational indicator of metal worker (yes/no) was evaluated as the possible risk factor in this study; results are:

		Controls	
		Yes	No
Cases	Yes	2	26
	No	5	12

(a) Find a 95 percent confidence interval for the odds ratio measuring the strength of the relationship between the disease and the exposure.

(b) Test for the independence between the disease and the exposure.

2. Ninety-eight heterosexual couples, at least one of whom was HIV-infected, were enrolled in an HIV transmission study and interviewed about sexual behavior (Padian, 1990). The following table provides a summary of condom use reported by heterosexual partners:

	Man		
Woman	Ever	Never	Total
Ever	45	6	51
Never	7	40	47
Total	52	46	98

Test to compare the reporting results between men and women.

3. A matched case-control study was conducted in order to evaluate the cumulative effects of acrylate and methacrylate vapors on olfactory function (Schwarts et al., 1989). Cases were defined as scoring at or below the 10th percentile on the UPSIT (University of Pennsylvania Smell Identification Test).

	Cases	
Controls	Exposed	Unexposed
Exposed	25	22
Unexposed	9	21

(a) Find a 95 percent confidence interval for the odds ratio measuring the strength of the relationship between the disease and the exposure.

(b) Test for the independence between the disease and the exposure.

4. A study in Maryland identified 4,032 white persons, enumerated in a nonofficial 1963 census, who became widowed between 1963 and 1974 (Helsing and Szklo, 1981). These people were matched, one-to-one, to married persons on the basis of race, sex, year of birth, and geography of residence. The matched pairs were followed to a second census in 1975, and we have the following overall male mortality test to compare the mortality of widowed men versus married men:

		Married Men	
		Died	Alive
Widowed	Died	2	292
Men	Alive	210	700

The data for 2,828 matched pairs of women were as follows, tested to compare the mortality of widowed women versus married women:

		Married Women	
		Died	Alive
Widowed	Died	1	264
Women	Alive	249	2314

5. Table 5.3 provides some data from a matched case-control study to investigate the association between the use of Xray and risk of childhood acute myeloidleukemia. In each matched set or pair, the case and control(s) were matched by age, race, and county of residence. The variables are:

- Matched set (or pair)—disease (1 = Case, 2 = Control)
- Some chracteristics of the child: Sex (1 = Male, 2 = Female), Downs syndrome (a known risk factor for leukemia; 1 = No, 2 = Yes), and Age
- Risk factors related to the use of Xray: MXray (Mother ever had Xray during pregnancy; 1 = No, 2 = Yes), UMXray (Mother ever had upper body Xray during pregnancy; 0 = No, 1 = Yes), LMXray (Mother ever had lower body Xray during pregnancy; 0 = No, 1 = Yes), FXray (Father ever had Xray; 1 = No, 2 = Yes), CXray (the Child ever had Xray; 1 = No, 2 = Yes), and CNXray (Child's total number of Xrays; 1 = None, 2 = 1 − 2, 3 = 3 − 4, 4 = 5 or more)

(a) Taken collectively, do the covariates contribute significantly to the separation of the cases and the controls?

(b) Fit the multiple regression model to obtain estimates of individual regression coefficients and their standard errors. Draw your conclusion concerning the conditional contribution of each factor.

(c) Within the context of the multiple regression model in (b), does sex alter the effect of Downs syndrome?

(d) Within the context of the multiple regression model in (b), taken collectively, does the exposure to Xray (by father, or mother, or child) relate significantly to the disease of the child?

(e) Within the context of the multiple regression model in (b), is the effect of age linear?

(f) Focusing on Downs syndrome (DS) as the primary factor, taken collectively, was this main effect altered by any other covariates?

TABLE 5.3. Xray-Leukemia Data

Matched Set	Disease	Sex	Downs	Age	MXRay	UMXRay	LMXay	FXRay	CXRay	CNXRay
1	1	2	1	0	1	0	0	1	1	1
	2	2	1	0	1	0	0	1	1	1
2	1	1	1	6	1	0	0	1	2	3
	2	1	1	6	1	0	0	1	2	2
3	1	2	1	8	1	0	0	1	1	1
	2	2	1	8	1	0	0	1	1	1
4	1	1	2	1	1	0	0	1	1	1
	2	1	1	1	1	0	0	1	1	1
5	1	1	1	4	2	0	1	1	1	1
	2	1	1	4	1	0	0	1	2	2
6	1	2	1	9	2	1	0	1	1	1
	2	1	1	9	1	0	0	1	1	1
7	1	2	1	17	1	0	0	1	2	2
	2	2	1	17	1	0	0	1	2	2
8	1	2	1	5	1	0	0	1	1	1
	2	1	1	5	1	0	0	1	1	1
9	1	2	2	0	1	0	0	1	1	1
	2	2	1	0	2	1	0	2	1	1
	2	2	1	0	1	0	0	1	1	1
10	1	2	1	7	1	0	0	2	1	1
	2	1	1	7	1	0	0	1	1	1
11	1	1	1	15	1	0	0	1	1	1
	2	1	1	15	1	0	0	1	2	2
12	1	1	1	12	1	0	0	1	2	2
	2	1	1	12	1	0	0	1	1	1
13	1	1	1	4	1	0	0	1	1	1
	2	2	1	4	1	0	0	1	1	1
14	1	1	1	14	1	0	0	1	2	2
	2	2	1	14	1	0	0	1	1	1
	2	1	1	14	1	0	0	1	1	1
15	1	1	1	7	1	0	0	2	1	1
	2	1	1	7	1	0	0	2	1	1
	2	1	1	7	1	0	0	1	2	2
16	1	1	1	8	1	0	0	1	2	2
	2	2	1	8	1	0	0	2	1	1
17	1	1	1	6	1	0	0	2	1	1
	2	1	1	6	1	0	0	1	1	1
18	1	2	1	13	1	0	0	2	2	3
	2	2	1	13	1	0	0	1	2	2
19	1	2	1	17	1	0	0	2	1	1
	2	1	1	17	1	0	0	1	2	2
20	1	2	1	5	2	0	1	1	2	4
	2	2	1	5	1	0	0	1	1	1
	2	1	1	5	1	0	0	1	1	1
21	1	1	1	13	1	0	0	1	2	4
	2	1	1	13	1	0	0	1	1	1
22	1	2	1	16	1	0	0	2	2	2
	2	2	1	16	1	0	0	2	1	1
	2	2	1	16	1	0	0	2	2	2
23	1	2	1	10	1	0	0	2	1	1
	2	2	1	10	1	0	0	1	1	1

TABLE 5.3. (*Continued*)

Matched Set	Disease	Sex	Downs	Age	MXRay	UMXRay	LMXay	FXRay	CXRay	CNXRay
24	1	1	1	0	1	0	0	2	1	1
	2	1	1	0	1	0	0	1	1	1
	2	2	1	0	2	0	1	1	1	1
25	1	1	1	1	1	0	0	2	1	1
	2	2	1	1	1	0	0	2	1	1
26	1	2	1	13	1	0	0	2	1	1
	2	1	1	13	2	1	1	1	1	1
27	1	1	1	11	1	0	0	1	1	1
	2	1	1	11	1	0	0	1	2	2
28	1	2	1	4	1	0	0	2	1	1
	2	2	1	4	1	0	0	1	2	2
29	1	2	1	1	1	0	0	2	1	1
	2	1	1	1	1	0	0	1	1	1
	2	2	1	1	1	0	0	1	1	1
30	1	2	1	15	2	0	1	2	2	3
	2	2	1	15	1	0	0	2	2	2
31	1	1	1	9	1	0	0	1	1	1
	2	2	1	9	1	0	0	1	1	1
32	1	1	1	15	1	0	0	2	2	2
	2	1	1	15	1	0	0	2	2	3
33	1	2	1	5	1	0	0	1	2	3
	2	2	1	5	2	1	0	1	1	1
34	1	1	1	10	2	0	1	2	1	1
	2	1	1	10	1	0	0	2	2	2
35	1	2	1	8	1	0	0	1	2	2
	2	2	1	8	1	0	0	1	1	1
36	1	2	1	15	1	0	0	1	2	4
	2	2	1	15	1	0	0	1	2	2
37	1	2	2	1	1	0	0	1	1	1
	2	2	1	1	2	1	0	2	1	1
38	1	2	1	0	1	0	0	1	1	1
	2	1	1	0	1	0	0	1	1	1
39	1	1	1	6	1	0	0	2	2	2
	2	2	1	6	1	0	0	1	1	1
40	1	1	1	14	1	0	0	2	2	2
	2	2	1	14	1	0	0	1	1	1
41	1	1	1	2	1	0	0	2	1	1
	2	2	1	2	1	0	0	1	1	1
42	1	1	2	1	2	1	0	1	1	1
	2	1	1	1	1	0	0	1	1	1
43	1	2	1	6	1	0	0	1	1	1
	2	2	1	6	1	0	0	1	1	1
44	1	1	1	16	1	0	0	1	1	1
	2	1	1	16	2	1	0	1	1	1
45	1	1	1	4	1	0	0	1	2	2
	2	2	1	4	1	0	0	1	1	1
46	1	2	1	1	1	0	0	1	1	1
	2	1	1	1	1	0	0	1	1	1
47	1	1	1	0	1	0	0	1	1	1
	2	1	1	0	1	0	0	2	1	1
48	1	2	1	0	1	0	0	1	1	1
	2	2	1	0	1	0	0	2	1	1

TABLE 5.3. (*Continued*)

Matched Set	Disease	Sex	Downs	Age	MXRay	UMXRay	LMXay	FXRay	CXRay	CNXRay
49	1	1	1	3	1	0	0	1	2	4
	2	1	1	3	1	0	0	1	2	4
50	1	2	1	5	1	0	0	1	1	1
	2	1	1	5	1	0	0	2	2	2
51	1	1	1	8	1	0	0	1	1	1
	2	1	1	8	1	0	0	1	1	1
52	1	2	1	9	1	0	0	2	2	2
	2	2	1	9	1	0	0	1	1	1
	2	2	1	9	1	0	0	1	2	2
53	1	1	1	2	1	0	0	1	1	1
	2	2	1	2	1	0	0	2	1	1
54	1	2	1	1	1	0	0	1	1	1
	2	2	1	1	2	1	0	1	1	1
55	1	1	1	3	1	0	0	1	2	2
	2	1	1	3	1	0	0	1	1	1
56	1	1	1	17	1	0	0	1	1	1
	2	1	1	17	2	1	0	1	2	2
	2	1	1	17	1	0	0	1	1	1
57	1	2	1	3	1	0	0	1	1	1
	2	2	1	3	1	0	0	2	1	1
58	1	1	1	10	1	0	0	1	2	4
	2	1	1	10	2	0	1	1	1	1
59	1	1	1	13	1	0	0	1	2	2
	2	2	1	13	2	0	1	1	1	1
60	1	1	1	0	1	0	0	1	1	1
	2	1	1	0	1	0	0	1	1	1
61	1	1	1	11	1	0	0	1	2	3
	2	1	1	11	1	0	0	2	1	1
62	1	1	1	14	1	0	0	1	2	2
	2	1	1	14	1	0	0	1	2	4
63	1	2	1	5	1	0	0	1	1	1
	2	1	1	5	1	0	0	1	1	1
	2	2	1	5	1	0	0	1	1	1
64	1	1	1	12	1	0	0	2	1	1
	2	2	1	12	1	0	0	1	1	1
65	1	2	1	9	2	0	1	1	2	2
	2	2	1	9	1	0	0	1	1	1
66	1	2	1	5	1	0	0	1	2	4
	2	1	1	5	1	0	0	2	1	1
67	1	1	1	8	1	0	0	1	1	1
	2	1	1	8	1	0	0	1	1	1
68	1	1	1	15	1	0	0	1	2	4
	2	1	1	15	1	0	0	2	1	1
	2	2	1	15	2	0	1	1	2	2
69	1	2	1	10	1	0	0	1	1	1
	2	2	1	10	1	0	0	1	1	1
70	1	1	1	3	1	0	0	1	1	1
	2	2	1	3	1	0	0	1	1	1
71	1	1	1	1	1	0	0	1	1	1
	2	1	1	1	1	0	0	1	1	1
	2	2	1	1	1	0	0	1	1	1
72	1	2	1	9	1	0	0	1	1	1
	2	2	1	9	1	0	0	2	1	1

TABLE 5.3. (*Continued*)

Matched Set	Disease	Sex	Downs	Age	MXRay	UMXRay	LMXay	FXRay	CXRay	CNXRay
73	1	1	1	1	1	0	0	1	1	1
	2	1	1	1	1	0	0	1	1	1
74	1	2	1	8	1	0	0	1	2	3
	2	2	1	8	1	0	0	2	2	2
75	1	1	1	12	1	0	0	1	1	1
	2	2	1	12	1	0	0	1	1	1
76	1	2	1	1	1	0	0	1	1	1
	2	2	1	1	1	0	0	1	1	1
77	1	2	1	4	1	0	0	1	1	1
	2	2	1	4	1	0	0	2	1	1
78	1	2	1	11	1	0	0	1	2	2
	2	2	1	11	1	0	0	1	1	1
79	1	1	1	2	1	0	0	1	1	1
	2	1	1	2	1	0	0	1	1	1
80	1	2	1	4	1	0	0	1	1	1
	2	2	1	4	1	0	0	1	1	1
81	1	1	1	1	1	0	0	2	1	1
	2	1	1	1	1	0	0	1	1	1
82	1	2	1	0	1	0	0	1	1	1
	2	1	1	0	1	0	0	1	1	1
83	1	1	1	5	1	0	0	1	1	1
	2	1	1	5	1	0	0	2	2	2
84	1	2	2	1	2	0	1	2	1	1
	2	1	1	1	1	0	0	2	1	1
85	1	1	1	12	1	0	0	2	2	2
	2	1	1	12	1	0	0	1	1	1
	2	1	1	12	1	0	0	2	1	1
86	1	1	1	12	2	0	1	2	2	2
	2	1	1	12	1	0	0	2	2	4
87	1	1	1	1	1	0	0	1	1	1
	2	1	1	1	1	0	0	1	1	1
88	1	1	1	9	1	0	0	2	1	1
	2	1	1	9	1	0	0	2	1	1
89	1	2	1	2	1	0	0	2	1	1
	2	2	1	2	1	0	0	1	1	1
90	2	2	1	1	1	0	0	1	1	1
	2	2	1	1	1	0	0	2	1	1
91	1	2	1	2	1	0	0	1	1	1
	2	2	1	2	1	0	0	1	1	1
92	1	1	1	15	1	0	0	2	1	1
	2	1	1	15	1	0	0	1	1	1
93	1	1	1	13	1	0	0	2	1	1
	2	1	1	13	1	0	0	1	1	1
94	1	1	1	6	1	0	0	2	2	4
	2	2	1	6	1	0	0	1	1	1
95	1	2	2	1	1	0	0	1	1	1
	2	2	1	1	1	0	0	1	1	1
96	1	2	1	8	1	0	0	1	1	1
	2	2	1	8	1	0	0	2	1	1
97	1	2	1	4	1	0	0	2	2	4
	2	2	1	4	1	0	0	2	1	1
98	1	1	1	6	1	0	0	1	2	2
	2	1	1	6	1	0	0	2	1	1

TABLE 5.3. (*Continued*)

Matched Set	Disease	Sex	Downs	Age	MXRay	UMXRay	LMXay	FXRay	CXRay	CNXRay
99	1	1	1	1	1	0	0	2	1	1
	2	2	1	1	1	0	0	1	1	1
100	1	1	1	14	1	0	0	1	2	4
	2	1	1	14	1	0	0	1	2	3
101	1	1	1	1	1	0	0	2	1	1
	2	1	1	1	1	0	0	1	1	1
102	1	2	2	1	1	0	0	1	1	1
	2	2	1	1	1	0	0	1	1	1
103	1	1	1	13	1	0	0	2	1	1
	2	1	1	13	1	0	0	1	1	1
104	1	2	1	13	1	0	0	1	2	4
	2	2	1	13	1	0	0	1	1	1
105	1	2	1	11	1	0	0	2	2	3
	2	2	1	11	1	0	0	1	1	1
106	1	2	1	13	1	0	0	1	2	3
	2	1	1	13	1	0	0	2	1	1
107	1	2	1	2	1	0	0	2	1	1
	2	2	1	2	1	0	0	1	1	1
108	1	2	1	7	1	0	0	1	1	1
	2	1	1	7	1	0	0	1	2	2
	2	2	1	7	1	0	0	1	2	2
109	1	2	1	16	1	0	0	1	2	3
	2	1	1	16	1	0	0	1	1	1
110	1	1	1	3	1	0	0	1	1	1
	2	1	1	3	1	0	0	1	1	1
111	1	1	1	13	1	0	0	2	2	4
	2	2	1	13	1	0	0	1	1	1
112	1	2	1	6	1	0	0	1	1	1
	2	2	1	6	1	0	0	1	1	1

CHAPTER 6

Methods for Count Data

Topics in Chapters 2 and 3 focused mainly on contingency tables and those in Chapter 4 on regression methods for binomial- and multinomial-distributed responses. This chapter is devoted to a different type of categorical data—*count data*; the eventual focus is the Poisson regression model. As usual, the purpose of the research is to assess relationships among a set of variables, one of which is taken to be the response or dependent variable, that is, a variable to be predicted from or explained by other variables called predictors, or explanatory or independent variables. Choosing an appropriate model and analytical technique depends on the type of response variable under investigation. The Poisson regression model applies when the dependent variable follows a Poisson distribution.

6.1. THE POISSON DISTRIBUTION

The binomial distribution of Chapter 2 is used to characterize an experiment when each trial of the experiment has two possible outcomes (often referred to as "failure" and "success"). Let the probabilities of failure and success be, respectively, $1-\pi$ and π, and we "code" these two outcomes as 0 (zero successes) and 1 (one success). The experiment consists of n repeated trials satisfying these assumptions:

(i) The n trials are all independent.
(ii) The parameter π is the same for each trial.

The target for the binomial distribution is the total number X of successes in n trials. The distribution of X is characterized by:

$$\Pr(X = x) = \binom{n}{x} \pi^x (1-\pi)^{n-x} \qquad \text{for} \quad x = 0, 1, 2, \ldots, n.$$

When n is large, direct calculation of binomial probabilities can involve a prohibitive amount of work. However, it can be shown that the limiting form of the binomial distribution, when $n \to \infty$, $\pi \to 0$ while $\theta = n\pi$ remains constant, is rather simple. It is given by

$$\begin{aligned} \Pr(X = x) &= \frac{\theta^x e^{-\theta}}{x!} \qquad \text{for} \quad x = 0, 1, 2, \ldots \\ &= p(x; \theta) \end{aligned}$$

and a random variable having this probability function is said to have the Poisson distribution $P(\theta)$. To illustrate the Poisson approximation of the binomial distribution, consider $n = 48$ and $p = .05$. Then

$$\begin{aligned} b(x = 5; n, p) &= .059 \\ p(x = 5; \theta) &= .060. \end{aligned}$$

The Poisson model is often used when the random variable X is supposed to represent the number of occurrences of some random event in an interval of time or space, or some volume of matter. Numerous applications in health sciences have been documented, for example, the number of viruses in a solution, the number of defective teeth per individual, the number of focal lesions in virology, the number of victims of specific diseases, the number of cancer deaths per household, the number of infant deaths in a certain locality during a given year, among others. The mean and variance of the Poisson distribution are equal:

$$\mu = \theta$$

$$\sigma^2 = \theta.$$

Given a sample of counts from the Poisson distribution $P(\theta)$, $\{x_i\}_{i=1}^n$, then, the sample mean $\bar{x}$ is an unbiased estimator for θ; its standard error is given by:

$$\mathrm{SE}(\bar{x}) = \sqrt{\frac{\bar{x}}{n}}.$$

Example 6.1. In estimating the infection rates in populations of organisms, sometimes it is impossible to assay each organism individually. Instead, the organisms are randomly divided into a number of pools and each pool is tested as a unit. Let

$N =$ number of insects in the sample
$n =$ number of pools used in the experiment
$m =$ number of insects per pool, $N = nm$ (for simplicity, assume that m is the same for every pool)

The random variable X concerned is the number of pools that show negative test results, (i.e., none of the insects are infected).

(i) Let λ be the population infection rate; the probability that all m insects in a pool are negative (in order to have a negative pool) is given by

$$\pi = (1 - \lambda)^m.$$

Designating a negative pool as "success," we have a binomial distribution for X which is $B(n, \pi)$.

(ii) In situations where the infection rate λ is very small, the Poisson distribution could be used as approximation with $\theta =$

$m\lambda$ being the expected number of infected insects in a pool. The Poisson probability corresponding to this number being zero is

$$\pi = \exp(-\theta),$$

and we have the same binomial distribution $B(n, \pi)$ for the experiment. For detailed description, see Bhattacharyya et al. (1979), Walter et al. (1980), or Le (1981), for extension of the above procedure to using pools of variable size.

It is interesting to note that testing for syphilis (as well as other very rare diseases) in the Army used to be done this way. The method was also adapted to estimate the infectious potential of blood containing human immunodeficiency virus (AIDS) (Connett et al., 1990).

Example 6.2. For the year 1981, the infant mortality rate (IMR) for the United States was 11.9 deaths per 1000 live births. For the same period, the New England states (Connecticut, Maine, Massachusetts, New Hampshire, Rhode Island, and Vermont) had 164,200 live births and 1,585 infant deaths (Freeman, 1980). If the national IMR applies, the mean and variance of the number of infant deaths in the New England states would be:

$$(164.2)(11.9) = 1{,}954.$$

From the z-score,

$$z = \frac{1{,}585 - 1{,}954}{\sqrt{1{,}954}}$$

$$= -8.35,$$

it is clear that the IMR in the New England states is below the national average.

Example 6.3. Cohort studies are designs in which one enrolls a group of healthy persons and follows them over certain periods of time; examples include occupational mortality studies. The cohort study design focuses attention on a particular exposure rather than a particular disease as in case-control studies. Advantages of a longitudinal approach include the opportunity for more accurate measurement of exposure history and a careful examination of the time relationships between exposure and disease.

The observed mortality of the cohort under investigation often needs to be compared with that expected from the death rates of the national population (served as standard), with allowance made for age, sex, race, and time period. Rates may be calculated either for total deaths or for separate causes of interest. The statistical method is often referred to as the *person-years* method. The basis of this method is the comparison of the observed number of deaths, d, from the cohort with the mortality that would have been expected if the group had experienced similar death rates to those of the standard population of which the cohort is a part. The expected number of deaths is usually calculated using published national life tables and the method is similar to that of indirect standardization of death rates.

Each member of the cohort contributes to the calculation of the expected deaths for those years in which he or she was at risk of dying during the study. There are three types of subjects:

(i) Subjects still alive on the analysis date.
(ii) Subjects who died on a known date within the study period.
(iii) Subjects who are lost to follow-up after a certain date. These cases are a potential source of bias; effort should be expended in reducing the number of subjects in this category.

Figure 6.1 shows the sitution illustrated by one subject of each type, from enrollment to the study termination (e.g., Le, 1997a).

Each subject is represented by a diagonal line that starts at the age and year at which the subject entered the study and continues as long as the subject is at risk of dying in the study.

In Figure 6.1, each subject is represented by a line starting from the year and age at entry and continuing until the study date, the date of death, or the date the subject was lost to follow-up. Period and age are divided into five-year intervals corresponding to the usual availability of referenced death rates. Then a quantity, r, is

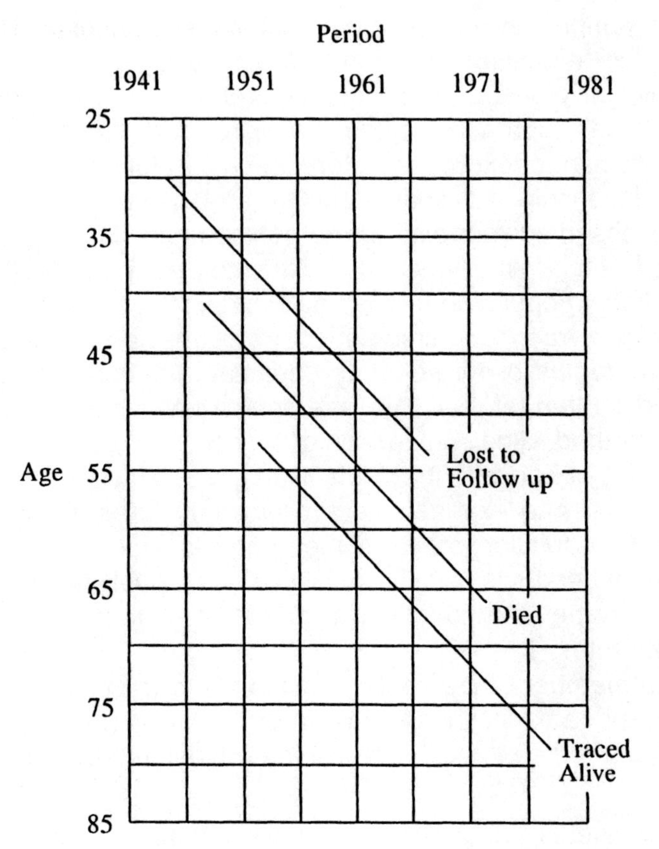

FIGURE 6.1. Representation of Basis of Subject-Years Method

defined for each individual as the cumulative risk over the follow-up period

$$r = \sum x\omega,$$

where summation is over each square in Figure 6.1 entered by the follow-up line, x is the time spent in a square, and ω the corresponding death rate for the given age-period combination. For the cohort, the individual values of r are added to give the expected number of deaths

$$m = \sum r.$$

For various statistical analyses, the observed number of deaths d may be treated as a Poisson variable with mean $\theta = m\rho$ where ρ is the relative risk of being a member of the cohort as compared to the standard population (Berry, 1983).

6.2. TESTING GOODNESS-OF-FIT

A goodness-of-fit test is used when one wishes to decide if an observed distribution of frequencies is incompatible with some hypothesized distribution. The Poisson is a very special distribution; its mean and its variance are equal. Therefore, given a sample of count data $\{x_i\}_{i=1}^{n}$, we often wish to know whether these data provide sufficient evidence to indicate that the sample did not come from a Poisson-distributed population. The hypotheses are:

H_0: The sampled population is distributed as Poisson.
H_A: The sampled population is not distributed as Poisson.

The most frequent violation is an overdispersion; the variance is larger than the mean. The implication is serious; the analysis which assumes the Poisson model often underestimates standard error(s), and thus wrongly inflates the level of significance.

The test statistic is the familiar Pearson's chi-square,

$$X^2 = \sum_{i}^{k} \frac{(O_i - E_i)^2}{E_i},$$

where O_i and E_i refer to the ith observed and expected frequencies, respectively (we used notations x_{ij} and $\widehat{m_{ij}}$ in Chapter 2). In this formula, k is the number of groups for which observed and expected frequencies are available. When the null hypothesis is true, the test statistic is distributed as chi-square with $(k - 2)$ degrees of freedom; one degree of freedom was lost because the mean needs to be estimated and one was lost because of the constraint $\sum O_i = \sum E_i$. It is also recommended that adjacent groups at the bottom of the table be combined in order to avoid having any expected frequencies less than 1.

TABLE 6.1. Emergency Service Data

No. of Complaints	Observed O_i	Expected E_i
0	1	1.54
1	12	5.19
2	12	8.71
3	5	9.68
4	1	8.10
5	4	5.46
6	2	2.99
7	2	1.45
8	2	0.62
9	1	0.22
10	1	0.09
11	1	0.04

Example 6.4. The purpose of this study was to examine the data for 44 physicians working for an emergency service at a major hospital. The response variable is the number of complaints received during the previous year; other details of the study are given in the next example. For the purpose of testing the goodness-of-fit, the data are summarized as in Table 6.1.

To obtain the expected frequencies, we first obtain relative frequencies by evaluating the Poisson probability for each value of $X = x$,

$$\Pr(X = x) = \frac{\theta^x e^{-\theta}}{x!} \quad \text{for} \quad x = 0, 1, 2, \ldots .$$

Since

$$\begin{aligned}\hat{\theta} &= \frac{\sum x_i}{44} \\ &= 3.34,\end{aligned}$$

we have, for example,

$$\begin{aligned}\Pr(X = 2) &= \frac{3.34^2 e^{-3.34}}{2!} \\ &= 0.198.\end{aligned}$$

Each of the expected relative frequencies is multiplied by the sample size, 44, to obtain the expected frequencies, e.g.,

$$\begin{aligned} E_2 &= (44)(0.198) \\ &= 8.71. \end{aligned}$$

In order to avoid having any expected frequencies less than 1, we combine the last five groups together, resulting in eight groups available for testing goodness-of-fit with

$$\begin{aligned} O_8 &= 2+2+1+1+1 \\ &= 7, \qquad \text{and} \\ E_8 &= 1.45+0.62+0.22+0.09+0.04 \\ &= 2.42. \end{aligned}$$

The result is

$$\begin{aligned} X^2 &= \frac{(1-1.59)^2}{1.19} + \frac{(12-5.19)^2}{5.19} + \cdots + \frac{(7-2.42)^2}{2.42} \\ &= 28.24, \end{aligned}$$

with $8-2=6$ degrees of freedom, indicating a significant deviation from the Poisson distribution ($p < 0.005$). A simple inspection of Table 6.1 reveals an obvious overdispersion.

6.3. THE POISSON REGRESSION MODEL

As previously mentioned, the Poisson model is often used when the random variable X is supposed to represent the number of occurrences of some random event in an interval of time or space, or some volume of matter, and numerous applications in health sciences have been documented. In some of these applications, one may be interested in to see if the Poisson-distributed dependent variable can be predicted from or explained by other variables. The other variables are called predictors, or explanatory or independent variables. For example, we may be interested in the number of defective teeth per individual as a function of gender and age of a child, brand of toothpaste, and whether the family has or does not have dental insurance. In this and other examples, the dependent variable Y is assumed to follow a Poisson distribution with mean θ. The Poisson regression model expresses this mean as a function of certain independent variables $X_1, X_2, \ldots, X_k$, in addition to the *size* of the observation unit from which one obtained the count of interest. For example, if Y is the number of viruses in a solution then the *size* is the volume of the solution; or if Y is the number of defective teeth then the *size* is the total number of teeth for that same individual. The following is the continuation of Example 6.4 on the emergency service data; but data in Table 6.2 also include information on four covariates.

Example 6.5. The purpose of this study was to examine the data for 44 physicians working for an emergency service at a major hospital so as to determine which of the following four variables are related to the number of complaints received during the previous year. In addition to the number of complaints, served as the dependent variable, data available consist of the number of visits—which serves as the *size* for the observation unit, the physician—and four covariates. Table 6.2 presents the comple data set. For each of the 44 physicians there are two continuous independent variables, the Revenue (dollars per hour) and work load at the emergency service (Hours), and two binary variables, Gender (Female/Male) and Residency training in emergency services (No/Yes).

6.3.1. Simple Regression Analysis

In this section we will discuss the basic ideas of simple regression analysis when only one predictor or independent variable is available for predicting the response of interest.

The Poisson Regression Model

In our framework, the dependent variable Y is assumed to follow a Poisson distribution; its values $y_i's$ are available from n "observation units," which is also characterized by an independent variable X. For the observation unit i $(1 \leq n)$, let s_i be the size and x_i be the covariate value. The Poisson regression model assumes that the relationship

TABLE 6.2. Emergency Service Data

N-Visits	Complaint	Gender	Residency	Revenue	Hours
2014	2	Y	F	263.03	1287.25
3091	3	N	M	334.94	1588.00
879	1	Y	M	206.42	705.25
1780	1	N	M	226.32	1005.50
3646	11	N	M	288.91	1667.25
2690	1	N	M	275.94	1517.75
1864	2	Y	M	295.71	967.00
2782	6	N	M	224.91	1609.25
3071	9	N	F	249.32	1747.75
1502	3	Y	M	269.00	906.25
2438	2	N	F	225.61	1787.75
2278	2	N	M	212.43	1480.50
2458	5	N	M	211.05	1733.50
2269	2	N	F	213.23	1847.25
2431	7	N	M	257.30	1433.00
3010	2	Y	M	326.49	1520.00
2234	5	Y	M	290.53	1404.75
2906	4	N	M	268.73	1608.50
2043	2	Y	M	231.61	1220.00
3022	7	N	M	241.04	1917.25
2123	5	N	F	238.65	1506.25
1029	1	Y	F	287.76	589.00
3003	3	Y	F	280.52	1552.75
2178	2	N	M	237.31	1518.00
2504	1	Y	F	218.70	1793.75
2211	1	N	F	250.01	1548.00
2338	6	Y	M	251.54	1446.00
3060	2	Y	M	270.52	1858.25
2302	1	N	M	247.31	1486.25
1486	1	Y	F	277.78	933.75
1863	1	Y	M	259.68	1168.25
1661	0	N	M	260.92	877.25
2008	2	N	M	240.22	1387.25
2138	2	N	M	217.49	1312.00
2556	5	N	M	250.31	1551.50
1451	3	Y	F	229.43	973.75
3328	3	Y	M	313.48	1638.25
2927	8	N	M	293.47	1668.25
2701	8	N	M	275.40	1652.75
2046	1	Y	M	289.56	1029.75
2548	2	Y	M	305.67	1127.00
2592	1	N	M	252.35	1547.25
2741	1	Y	F	276.86	1499.25
3763	10	Y	M	308.84	1747.50

between the mean of Y and the covariate X is described by

$$\begin{aligned} E(Y_i) &= s_i \lambda(x_i) \\ &= s_i \exp(\beta_0 + \beta_1 x_i), \end{aligned}$$

where $\lambda(x_i)$ is called the *risk* of observation unit i $(1 \leq n)$.

Under the assumption that Y_i is Poisson, the likelihood function is given by

$$L(y;\beta) = \prod_{i=1}^{n} \left\{ \frac{[s_i \lambda(x_i)]^{y_i} \exp[-s_i \lambda(x_i)]}{y_i!} \right\}$$

or

$$\ln L(y;\beta) = \sum_{i=1}^{n} \{y_i \ln s_i - \ln y_i! + y_i[\beta_0 + \beta_1 x_i] - s_i \exp[\theta_0 + \beta_1 x_i]\},$$

from which estimates for β_0 and β_1 can be obtained by the maximum likelihood procedure.

Measure of Association

Consider the case of a binary covariate X, representing an exposure (1 = exposed, 0 = not exposed). We have the following:

(i) If the observation unit is exposed, then

$$\ln \lambda_i(\text{exposed}) = \beta_0 + \beta_1 + \ln s_i,$$

whereas

(ii) If the observation unit is not exposed, then

$$\ln \lambda_i(\text{not exposed}) = \beta_0 + \ln s_i,$$

or, in other words,

$$\frac{\lambda_i(\text{exposed})}{\lambda_i(\text{not exposed})} = e^{\beta_1}.$$

This quantity is called the relative risk associated with the exposure.

Similarly, we have for a continuous covariate X and any value x of X,

$$\ln \lambda_i(X = x) = \beta_0 + \beta_1 x + \ln s_i$$
$$\ln \lambda_i(X = x + 1) = \beta_0 + \beta_1 (x + 1) + \ln s_i$$

so that

$$\frac{\lambda_i(X = x + 1)}{\lambda_i(X = x)} = e^{\beta_1},$$

representing the relative risk associated with *one unit increase* in the value of X.

The basic rationale for using the terms *risk* and *relative risk* is the approximation of the binomial distribution by the Poisson distribution. Recall from the previous section that, when $n \to \infty$, $\pi \to 0$ while $\theta = n\pi$ remains constant, the binomial distribution $B(n, \pi)$ can be approximated by the Poisson distribution $P(\theta)$. The number n is the size of the observation unit; so the ratio between the mean and the size represents the π(or λ(x) in the new model), the probability or *risk* and the ratio of risks is the risk ratio or *relative risk*.

Example 6.6. Refer to the emergency service data in Example 6.5 and suppose we want to investigate the relationship between the number of complaints (adjusted for number of visits) and residency training. It may be perceived that by having training in the specialty a physician would perform better and, therefore, be less likely to provoke complaints. An application of the simple Poisson regression analysis yields:

Variable	Coefficient	St. Error	z Statistic	p-Value
Intercept	−6.7566	0.1387	−48.714	< 0.0001
No Residency	0.3041	0.1725	1.763	0.0779

The result indicates that the common perception is almost true, that the relationship between the number of complaints and no residency training in emergency service is marginally significant ($p = 0.0779$); the relative risk associated with no residency training is

$$\exp(.3041) = 1.36.$$

Those without previous training are 36 percent more likely to receive the same number of complaints as those who were trained in the specialty.

Note: An SAS program would include these instructions:

```
DATA EMERGENCY;
INPUT VISITS CASES RESIDENCY;
LN=LOG(VISITS);
CARDS;
(Data);
PROC GENMOD DATA EMERGENCY;
CLASS RESIDENCY;
MODEL CASES=RESIDENCY/DIST=POISSON LINK=LOG
OFFSET=LN;
```

where EMERGENCY is the name assigned to the data set, VISITS is the number of visits, CASES is the number of complaints, and RESIDENCY is the binary covariate indicating whether the physician received residency training in the specialty. The option CLASS is used to declare that the covariate is categorical.

6.3.2. Multiple Regression Analysis

The effect of some factor on a dependent or response variable may be influenced by the presence of other factors through effect modifications (i.e., interactions). Therefore, in order to provide a more comprehensive analysis, it is very desirable to consider a large number of factors and sort out which ones are most closely related to the dependent variable. This method, which is multiple Poisson regression analysis, involves a linear combination of the explanatory or independent variables; the variables must be quantitative with particular numerical values for each observation unit. A covariate or independent variable may be dichotomous, polytomous, or continuous; categorical factors will be represented by dummy variables. In many cases, data transformations of continuous measurements (e.g., taking the logarithm) may be desirable so as to satisfy the linearity assumption.

Poisson Regression Model with Several Covariates

Suppose we want to consider k covariates, $X_1, X_2, \ldots, X_k$, simultaneously. The simple Poisson regression model of the previous section can be easily generalized and expressed as

$$\begin{aligned} E(Y_i) &= s_i \lambda(x'_{ji}\ s) \\ &= s_i \exp\left(\beta_0 + \sum_{j=1}^{k} \beta_j x_{ji}\right), \end{aligned}$$

where Y is the Poisson-distributed dependent variable and $\lambda(x'_{ji}\ s)$ is the *risk* of observation unit i $(1 \leq n)$.

Under the assumption that Y_i is Poisson, the likelihood function is given by

$$L(y;\beta) = \prod_{i=1}^{n} \left\{ \frac{[s_i \lambda(x_i)]^{y_i} \exp[-s_i \lambda(x'_{ji}\ s)]}{y_i!} \right\}$$

or

$$\begin{aligned} \ln L(y;\beta) = \sum_{i=1}^{n} \Bigg\{ & y_i \ln s_i - \ln y_i! + y_i \left[\beta_0 + \sum_{j=1}^{k} \beta_j x_{ji}\right] \\ & - s_i \exp\left[\theta_0 + \sum_{j=1}^{k} \beta_j x_{ji}\right] \Bigg\}, \end{aligned}$$

from which estimates for $\beta_0, \beta_1, \ldots, \beta_k$ can be obtained by the maximum likelihood procedure.

Also similar to the simple regression case, $\exp(\beta_i)$ represents:

(i) the relative risk associated with an exposure if X_i is binary (exposed $X_i = 1$ vs. unexposed $X_i = 0$), or

(ii) the relative risk due to one unit increase if X_i is continuous ($X_i = x + 1$ vs. $X_i = x$), and after $\hat{\beta}_i$ and its standard error have been obtained, a 95 percent confidence interval for the above relative risk is given by:

$$\exp[\hat{\beta}_i \pm 1.96\mathrm{SE}(\hat{\beta}_i)].$$

These results are necessary in the effort to identify important risk factors for the Poisson outcome, the *count*. Of course, before such analyses are done, the problem and the data have to be examined carefully. If some of the variables are highly correlated, then one or fewer of the correlated factors are likely to be as good predictors as all of them; information from other similar studies also has to be incorporated so as to drop some of these correlated explanatory variables. The uses of products, such as X_1X_2, and higher-power terms, such as X_1^2, may be necessary and can improve the goodness-of-fit. It is important to note that we are assuming a *(log)linear* regression model in which, for example, the relative risk due to one unit increase in the value of a continuous X_i ($X_i = x + 1$ vs. $X_i = x$) is independent of x. Therefore, if this *linearity* seems to be violated, the incorporation of powers of X_i should be seriously considered. The use of products will help in the investigation of possible effect modifications. And, finally, there is the messy problem of missing data; most packaged programs would delete a subject if one or more covariate values are missing.

Testing Hypotheses in Multiple Poisson Regression

Once we have fit a multiple Poisson regression model and obtained estimates for the various parameters of interest, we want to answer questions about the contributions of various factors to the prediction of the Poisson-distributed response variable. There are three types of such questions:

(i) An overall test: Taken collectively, does the entire set of explanatory or independent variables contribute significantly to the prediction of response?

(ii) Test for the value of a single factor: Does the addition of one particular variable of interest add significantly to the prediction of response over and above that achieved by other independent variables?

(iii) Test for contribution of a group of variables: Does the addition of a group of variables add significantly to the prediction of response over and above that achieved by other independent variables?

Overall Regression Test We now consider the first question stated above concerning an overall test for a model containing k factors. The null hypothesis for this test may stated as: "all k independent variables *considered together* do not explain the variation in the response any more than the size alone." In other words,

$$\mathcal{H}_0 : \beta_1 = \beta_2 = \cdots = \beta_k = 0.$$

This can be tested using the likelihood ratio chi-square test at k degrees of freedom,

$$\chi^2 = 2[\ln L_k - \ln L_0],$$

where $\ln L_k$ is the log-likelihood value for the model containing all k covariates and $\ln L_0$ is the log-likelihood value for the model containing only the intercept. Computer packaged programs, such as SAS PROC GENMOD provide these log-likelihood values, but in separate runs. □

Example 6.7. Refer to the data set on emergency service of Example 6.5 with four covariates: Gender, Residency, Revenue, and work load (Hours). We have:

(i) with all four covariates included, $\ln L_4 = 47.783$, whereas
(ii) with no covariates included, $\ln L_0 = 43.324$,

leading to $\chi^2 = 8.918$ with 4 dfs ($p < .05$), indicating that at least one covariate must be significantly related to the number of complaints.

Note: For model (i), the SAS program would include this instruction:

```
MODEL CASES=GENDER RESIDENCY REVENUE HOURS/
DIST=POISSON LINK=LOG OFFSET=LN;
```

and for model (ii);

```
MODEL CASES=/ DIST=POISSON LINK=LOG
OFFSET=LN;
```

(See note after Example 6.6 for other details of the program.)

Test for a Single Variable Let us assume that we now wish to test whether the addition of one particular independent variable of interest adds significantly to the prediction of the response over and above that achieved by other factors already present in the model. The null hypothesis for this test may stated as: "Factor X_i does not have any value added to the prediction of the response *given that other factors are already included in the model.*" In other words,

$$\mathcal{H}_0 : \beta_i = 0.$$

To test such a null hypothesis, one can use

$$z_i = \frac{\hat{\beta}_i}{\mathrm{SE}(\hat{\beta}_i)},$$

where $\hat{\beta}_i$ is the corresponding estimated regression coefficient and $\mathrm{SE}(\hat{\beta}_i)$ is the estimate of the standard error of $\hat{\beta}_i$, both of which are printed by standard computer packaged programs such as SAS. In performing this test, we refer the value of the z score to percentiles of the standard normal distribution; for example, we compare the absolute value of z to 1.96 for a two-sided test at the 5 percent level. □

Example 6.8. Referring to the data set on emergency service of Example 6.5 with all four covariates, we have:

Variable	Coefficient	St. Error	z Statistic	p-Value
Intercept	−8.1338	0.9220	−8.822	< .0001
No Residency	0.2090	0.2012	1.039	0.2988
Female	−0.1954	0.2182	−0.896	0.3703
Revenue	0.0016	0.0028	0.571	0.5775
Hours	0.0007	0.0004	1.750	0.0452

Only the effect of work load (Hours) is significant at the 5 percent level.

Note: Use the same SAS program as in Examples 6.6 and 6.7.

Given a continuous variable of interest, one can fit a polynomial model and use this type of test to check for linearity (see *type 1 analysis* in the next section). It can also be used to check for a single product representing an effect modification.

Example 6.9. Refer to the data set on emergency service of Example 6.5, but this time we investigate only one covariate, the work load (Hours). After fitting the second-degree polinomial model,

$$E(Y_i) = s_i \exp(\beta_0 + \beta_1 \text{Hour}_i + \beta_2 \text{Hour}_i^2),$$

we obtained a result which indicates that the *curvature effect* is negligible ($p = 0.8797$).

Note: An SAS program would include the instruction:

```
MODEL CASES=HOURS HOURS*HOURS/ DIST=POISSON
LINK=LOG OFFSET=LN;
```

The following is another interesting example comparing the incidences of nonmelanoma skin cancer among women from two major metropolitan areas, one in the South and one in the North.

Example 6.10. In this example, the dependent variable is the number of cases of skin cancer. Data were obtained from two metropolitan areas: Minneapolis–St. Paul and Dallas–Ft. Worth. The population of each area is divided into eight age groups and the data are shown in Table 6.3 (Kleinbaum et al., 1988). This problem involves two covariates: age and location. Both are categorical. Using seven dummy variables to represent the eight age groups (with 85+ being the baseline) and one for location (with Minneapolis–St. Paul as the baseline), we obtain the results in the following table. These results indicate a clear upward trend of skin cancer incidence with age and, with Minneapolis–St. Paul as the baseline,

TABLE 6.3. Skin Cancer Data

City:	Minneapolis–St. Paul		Dallas–Ft. Worth	
Age Group	Cases	Population	Cases	Population
15–24	1	172,675	4	181,343
25–34	16	123,065	38	146,207
35–44	30	96,216	119	121,374
45–54	71	92,051	221	111,353
55–64	102	72,159	259	83,004
65–74	130	54,722	310	55,932
75–84	133	32,185	226	29,007
85+	40	8,328	65	7,538

the relative risk associated with Dallas–Ft.Worth is

$$\begin{aligned}\text{RR} &= \exp(0.8043) \\ &= 2.235,\end{aligned}$$

Variable	Coefficient	St. Error	z Statistic	p-Value
Intercept	−5.4797	0.1037	52.842	< 0.0001
Age 15–24	−6.1782	0.4577	−13.498	< 0.0001
Age 25–34	−3.5480	0.1675	−21.182	< 0.0001
Age 35–44	−2.3308	0.1275	−18.281	< 0.0001
Age 45–54	−1.5830	0.1138	−13.910	< 0.0001
Age 55–64	−1.0909	0.1109	−9.837	< 0.0001
Age 65–74	−0.5328	0.1086	−4.906	< 0.0001
Age 75–84	−0.1196	0.1109	−1.078	0.2809
Dallas–Ft.Worth	0.8043	0.0522	15.408	< 0.0001

an increase of more than twofold for this southern metropolitan area.

Note: An SAS program would include the instruction:

```
INPUT AGEGROUP CITY $ POP CASES;
LN=LOG(POP);
MODEL CASES=AGEGROUP CITY/ DIST=POISSON
LINK=LOG OFFSET=LN;
```

Contribution of a Group of Variables This testing procedure addresses the more general problem of assessing the additional contribution of two or more factors to the prediction of the response over and above that made by other variables already in the regression model. In other words, the null hypothesis is of the form

$$\mathcal{H}_0 : \beta_1 = \beta_2 = \cdots = \beta_m = 0.$$

To test such a null hypothesis, one can perform a likelihood ratio chi-squared test, with m dfs,

$$\chi^2_{\text{LR}} = 2[\ln L(\hat{\beta};\ \text{all } X\text{'s}) - \ln L(\hat{\beta};\ \text{all other } X\text{'s with } X\text{'s under investigation deleted})]. \qquad \square$$

As with the above *z test*, this *multiple contribution* procedure is very useful for assessing the importance of potential explanatory variables. In particular, it is often used to test whether a similar group of variables, such as *demographic characteristics*, is important for the prediction of the response; these variables have some trait in common. Another application would be a collection of powers *and/or* product terms (referred to as interaction variables). It is often of interest to assess the interaction effects collectively before trying to consider individual interaction terms in a model as previously suggested. In fact, such use reduces the total number of tests to be performed and this, in turn, helps to provide better control of overall Type I error rates which may be inflated due to multiple testing.

Example 6.11. Refer to the data set on skin cancer of Example 6.10 with all eight covariates. We consider, collectively, the seven dummy variables representing age; the basic idea is to see if there are any differences without drawing seven separate conclusions comparing each age group versus the baseline.

1. With all eight variables included, we obtained: $\ln L = 7201.864$.
2. When the seven age variables were deleted, we obtained: $\ln L = 5921.076$.

Therefore:

$$\chi^2_{LR} = 2[\ln L(\hat{\beta};\ \text{eight variables}) - \ln L(\hat{\beta};\ \text{only location variable})]$$

$$= 2561.568; \qquad 7 \text{ dfs}, \qquad p\text{-value} < .0001.$$

In other words, the difference between the age group is highly significant; in fact, it is more so than the difference between the cities.

Main Effects The z-tests for single variables are sufficient for investigating the effects of continuous and binary covariates. For categorical factors with several categories, such as the age group in the skin cancer data of Example 6.10, this process in PROC GENMOD would choose a *baseline* category and compare each other category versus the chosen baseline category. However, it is usually of interest to determine the importance of the *main effects* (i.e., one statistical test for each covariate, not each category of a covariate). This can be achieved using PROC GENMOD by two different ways: (i) treating the several category-specific effects as a *group* as seen in Example 6.11; this would requires two sepate computer runs, or (ii) requesting the *type 3 analysis* option as shown in the following example. □

Example 6.12. Referring to the skin cancer data of Example 6.10, the type 3 analysis shows:

Source	df	LR Chi-square	p-Value
Age Group	7	2561.57	< 0.0001
City	1	258.72	< 0.0001

The result for Age Group main effect is identical to that of Example 6.11.

Note: An SAS program would include the instruction:

```
MODEL CASES=AGEGROUP CITY/ DIST=POISSON
LINK=LOG OFFSET=LN TYPE3;
```

Specific and Sequential Adjustments In the type 3 analysis, or any other multiple regression analysis, we test the significance of the effect of each factor *added* to the model containing *all other factors*; that is, we investigate the *additional* contribution of the factor to the explanation of the dependent variable. Sometimes, however, we may be interested in a hierarchical or sequential adjustment. For example, we have Poisson-distributed response Y and three covariates X_1, X_2, and X_3; we want to investigate the effect of X_1 on Y (unadjusted), the effect of X_2 added to the model containing X_1, and the effect of X_3 added to the model containing X_1 and X_2. This can be achieved using PROC GENMOD by requesting the *type 1 analysis* option. □

Example 6.13. Referring to the data set on emergency service of Example 6.5,

(i) Type 3 analysis shows:

Source	df	LR Chi-square	p-Value
Residency	1	1.09	0.2959
Gender	1	0.82	0.3641
Revenue	1	0.31	0.5781
Hours	1	4.18	0.0409

(ii) Type 1 analysis shows:

Source	df	LR Chi-square	p-Value
Residency	1	3.199	0.0741
Gender	1	0.84	0.3599
Revenue	1	0.71	0.3997
Hours	1	4.18	0.0409

The results for physician hours are identical because they are adjusted for all three other covariates in both types of analysis. However, the results for other covariates are very different. The effect of Residency is marginally significant in type 1 analysis ($p = 0.0741$, unadjusted) and is not significant in type 3 analysis after adjusting for the other three covariates. Similarly, the results for Revenue are also different; in type 1 analysis they are adjusted only for Residency and Gender ($p = 0.3997$; the ordering of variables is specified in the INPUT statement of the computer program) whereas in type 3 analysis it is adjusted for all three other covariates ($p = 0.5781$).

Note: An SAS program would include the instruction:

```
MODEL CASES=RESIDENCY GENDER REVENUE HOURS/
DIST=POISSON LINK=LOG OFFSET=LN TYPE1 TYPE3;
```

6.3.3. Overdispersion

The Poisson is a very special distribution; its mean μ and its variance σ^2 are equal. If we use the variance-mean ratio as a dispersion parameter then it is 1 in a standard Poisson model, less than 1 in an underdispersed model, and greater than 1 in an overdispersed model. Overdispersion is a common phenomenon in practice and it causes concerns because the implication is serious; the analysis which assumes the Poisson model often underestimates standard error(s), and thus wrongly inflates the level of significance.

Measuring and Monitoring Dispersion

After a Poisson regression model is fitted, dispersion is measured by the scaled deviance or scaled Peason chi-square; it is the deviance or Pearson chi-square divided by the degrees of freedom (number of observations minus number of parameters). The deviance is defined as twice the difference between the maximum achievable log-likelihood and the log-likelihood at the maximum likelihood estimates of the regression parameters. The contribution to the Pearson chi-square from the ith observation is:

$$\frac{(y_i - \hat{\mu}_i)^2}{\hat{\mu}_i}$$

(Frome, 1983; Frome and Checkoway, 1985).

Example 6.14. Referring to the data set on emergency service of Example 6.5 with all four covariates, we have:

Criterion	df	Value	Scaled Value
Deviance	39	54.518	1.398
Pearson Chi-square	39	54.417	1.370

Both indices are greater than 1, indicating an overdispersion. In this example, we have a sample size of 44 but five degrees of freedom are lost due to the estimation of the five regression parameters, including the intercept.

Fitting an Overdispersed Poisson Model

PROC GENMOD allows the specification of a scale parameter to fit overdispersed Poisson regression models. The GENMOD procedure does not use the Poisson density function; it fits generalized linear models of which the Poisson model is a special case. Instead of

$$\text{Var}(Y) = \mu,$$

it allows the variance function to have a multiplicative overdispersion factor ϕ:

$$\text{Var}(Y) = \phi\mu.$$

The models are fitted in the usual way, and the point estimates of regression coefficient are not affected. The covariance matrix, however, is multiplied by ϕ. There are two options available for fitting overdispersed models; the users can control either the scaled deviance (by specifying DSCALE in the model statement) or the scaled Pearson chi-square (by specifying PSCALE in the model statement). The value of the controlled index becomes 1; the value of the other is close to but may not be equal to 1.

Example 6.15. Refer to the data set on emergency service of Example 6.5 with all four covariates; by fitting an overdispersed model controlling the scaled deviance, we have:

Variable	Coefficient	St. Error	z Statistic	p-Value
Intercept	−8.1338	1.0901	−7.462	< .0001
No Residency	0.2090	0.2378	0.879	0.3795
Female	−0.1954	0.2579	−0.758	0.4486
Revenue	0.0016	0.0033	0.485	0.6375
Hours	0.0007	0.0004	1.694	0.0903

As compared to the results in Example 6.8, the point estimates remain the same but the standard errors are larger; the effect of work load (Hours) is no longer significant at the 5 percent level.

Note: An SAS program would include this instruction:

```
MODEL CASES=GENDER RESIDENCY REVENUE HOURS/
DIST=POISSON LINK=LOG OFFSET=LN DSCALE;
```

and the measures of dispersion become:

Criterion	df	Value	Scaled Value
Deviance	39	39.000	1.000
Pearson Chi-square	39	38.223	0.980

We would obtain similar results by controlling the scaled Pearson chi-square.

6.3.4. Stepwise Regression

In many applications, our major interest is to identify important risk factors. In other words, we wish to identify from many available factors a small subset of factors that relate significantly to the outcome (e.g., the disease under investigation). In that identification process, of course, we wish to avoid a large Type I (false positive) error. In a regression analysis, a Type I error corresponds to including a predictor that has no real relationship to the outcome; such an inclusion can greatly confuse the interpretation of the regression results. In a standard multiple regression analysis, this goal can be achieved by using a strategy that adds into or removes from a regression model one factor at a time according to a certain order of relative importance. Therefore the two important steps are:

1. Specifying a criterion or criteria for selecting a model.
2. Specifying a strategy for applying the chosen criterion or criteria.

Strategies: This is concerned with specifying the strategy for selecting variables. Traditionally, such a strategy is concerned with whether a particular variable should be added to a model or whether any particular variable should be deleted from a model at a particular stage of the process. As computers became more accessible and more powerful, these practices became more popular.

Forward Selection Procedure In the forward selection procedure, we proceed as follows:

Step 1: Fit a simple logistic linear regression model to each factor, one at a time.

Step 2: Select the most important factor according to a certain predetermined criterion.

Step 3: Test for the significance of the factor selected in step 2 and determine, according to a certain predetermined criterion, whether or not to add this factor to the model.

Step 4: Repeat steps 2 and 3 for those variables not yet in the model. At any subsequent step, if none meets the criterion in step 3, no more variables are included in the model and the process is terminated. □

Backward Elimination Procedure In the backward elimination procedure, we proceed as follows:

Step 1: Fit the multiple logistic regression model containing all available independent variables.

Step 2: Select the least important factor according to a certain predetermined criterion; this is done by considering one factor at a time and treating it as though it were the last variable to enter.

Step 3: Test for the significance of the factor selected in step 2 and determine, according to a certain predetermined criterion, whether or not to delete this factor from the model.

Step 4: Repeat steps 2 and 3 for those variables still in the model. At any subsequent step, if none meets the criterion in step 3, no more variables are removed from the model and the process is terminated. □

Stepwise Regression Procedure Stepwise regression is a modified version of forward regression that permits reexamination, at every step, of the variables incorporated in the model in previous steps. A variable entered at an early stage may become superfluous at a later stage because of its relationship with other variables now in the model. The information it provides becomes redundant. That variable may be removed, if meeting the elimination criterion, and the model is re-fitted with the remaining variables, and the forward process goes on. The whole process, one step forward followed by one step backward, continues until no more variables can be added or removed. Without an automatic computer algorithm, this comprehensive strategy may be too tedious to implement.

□

Criteria: For the first step of the forward selection procedure, decisions are based on individual score test results (chi-squared, 1df). In subsequent steps, both forward and backward, the ordering of levels of importance (step 2) and the selection (test in step 3) are based on the likelihood ratio chi-squared statistic:

$$\chi^2_{\text{LR}} = 2[\ln L(\hat{\beta};\text{ all other } X\text{'s}) - \ln L(\hat{\beta};\text{ all other } X\text{'s with one } X \text{ deleted})].$$

In the case of Poisson regression, computer packaged programs, such as SAS PROC GENMOD, do not have an automatic stepwise option. Therefore, the implementation is much more tedious and time-consuming. In selecting the first variable (step 1), we have to fit simple regression models to each and every factor separately, then decide, based on the computer output, on the first selection before coming back for computer runs in step 2. At subsequent steps, we can tave advantage of *type 1 analysis* results.

Example 6.16. Refer to the data set on emergency service of Example 6.5 with all four covariates: work load (Hours), Residency, Gender, and Revenue. This time we perform a regression analysis using forward selection in which we specify that a variable has to be significant at the 0.10 level before it can enter into the model. In addition, we fit all overdispersed models using the DSCALE option in PROC GENMOD.

The results of the four simple regression analyses are:

Variable	LR χ^2	p-Value
Hours	4.136	0.0420
Residency	2.166	0.1411
Gender	0.845	0.3581
Revenue	0.071	0.7897

Work load (Hours) meets the entrance criterion and is selected. In the next step, we fit three models each with two covariates: Hours and Residency, Hours and Gender, and Hours and Revenue. The following table shows the significance of each added variable to the model containg Hours using type 1 analysis:

Variable	LR χ^2	p-Value
Residency	0.817	0.3662
Gender	1.273	0.2593
Revenue	0.155	0.6938

None of these three variables meets the .10 level for entry into the model.

6.4. EXERCISES

Inflammation of the middle ear, or *otitis media* (OM), is one of the most common childhood illnesses and accounts for one-third of the practice of pediatrics during the first five years of life. Understanding the natural history of otitis media is of considerable importance due to morbidity for the child as well as concern about long-term effects on behavior, speech, and language development. In an attempt to understand that natural history, large groups of pregnant women were enrolled and their newborns were followed from birth. The response variable is the number of episodes of otitis media in the first six months (NBER) and potential factors under investigation are upper repiratory infection (URI), sibling history of otitis media (SIBHX; 1 for yes), daycare, number of cigarettes consumed per day by the parents (CIGS), cotinin level (CNIN) measured from the urine of the baby (some marker for exposure to cigarette smoke), and whether the baby was born in the fall season (FALL). Table 6.4 provides about half of our data set.

(a) Taken collectively, do the covariates contribute significantly to the prediction of the number of otitis media in the first six months?

(b) Fit the multiple regression model to obtain estimates of individual regression coefficients and their standard errors. Draw your conclusion concerning the conditional contribution of each factor.

(c) Is there any indication of overdispersion? If so, fit an appropriate overdispersed model and compare the results to those in (b).

(d) Refit the model (b) to implement this sequential adjustment:

$$\text{SIBHX} \rightarrow \text{DAYCARE} \rightarrow \text{CIGS} \rightarrow \text{CNIN} \rightarrow \text{FALL}$$

(e) Within the context of the multiple regression model in (b), does daycare alter the effect of sibling history?

(f) Within the context of the multiple regression model in (b), is the effect of cotinin level linear?

(g) Focus on sibling history of otitis media (SIBHX) as the primary factor; taken collectively, was this main effect altered by any other covariates?

TABLE 6.4. Otitis Media Data

UIR	SIBHX	DAYCARE	CIGS	CNIN	FALL	NBER
1	0	0	0	0.00	0	2
1	0	0	0	27.52	0	3
1	0	1	0	0.00	0	0
1	0	0	0	0.00	0	0
0	1	1	0	0.00	0	0
1	0	0	0	0.00	0	0
1	0	1	0	0.00	0	0
0	0	1	0	0.00	0	0
0	1	0	8	83.33	0	0
1	0	1	0	89.29	0	0
0	0	1	0	0.00	0	1
0	1	1	0	32.05	0	0
1	0	0	0	471.40	0	0
1	0	0	0	0.00	0	1
1	0	1	0	12.10	0	0
0	1	0	5	26.64	0	0
0	0	1	0	40.00	0	0
1	0	1	0	512.05	0	0
0	0	1	0	77.59	0	0
1	0	1	0	0.00	0	0
1	0	1	0	0.00	0	0
0	0	1	0	0.00	0	0
1	0	1	0	0.00	0	3
1	0	0	0	0.00	0	1
0	0	1	0	21.13	0	0
1	0	0	0	15.96	0	1
1	0	0	0	0.00	0	1
0	0	0	0	0.00	0	0
1	0	0	0	9.26	0	0
1	0	1	0	0.00	0	0
1	0	0	0	0.00	0	1
0	0	0	0	0.00	0	0
0	1	0	0	0.00	0	0
0	0	1	0	0.00	0	0
1	0	1	0	525.00	0	2
0	0	0	0	0.00	0	0
1	0	1	0	0.00	0	0
1	0	1	0	0.00	0	1
1	0	0	0	0.00	0	0
0	0	0	0	57.14	0	0

TABLE 6.4. (*Continued*)

UIR	SIBHX	DAYCARE	CIGS	CNIN	FALL	NBER
1	1	1	0	125.00	0	0
1	0	0	0	0.00	0	0
0	1	0	0	0.00	0	0
1	0	1	0	0.00	0	3
1	0	1	0	0.00	0	0
0	0	0	0	0.00	0	1
1	0	1	0	0.00	0	1
1	0	1	0	80.25	0	2
0	0	0	0	0.00	0	0
1	0	1	0	219.51	0	1
1	0	0	0	0.00	0	0
1	0	0	0	0.00	0	1
0	0	0	0	0.00	0	0
1	0	0	0	0.00	0	0
1	0	1	0	0.00	0	2
1	0	1	0	8.33	0	2
1	0	0	0	12.02	0	5
0	1	1	40	297.98	0	0
1	0	1	0	13.33	0	0
1	0	0	0	0.00	0	3
1	1	1	25	110.31	0	3
1	0	1	0	0.00	0	1
1	0	1	0	0.00	0	1
1	0	0	0	0.00	0	1
0	0	0	0	0.00	0	0
0	0	0	0	0.00	0	0
1	0	1	0	0.00	0	1
1	0	1	0	0.00	0	1
1	0	0	0	285.28	0	1
1	0	0	0	0.00	0	1
1	0	0	0	15.00	0	2
1	0	0	0	0.00	0	0
1	0	0	0	13.40	0	1
1	0	1	0	0.00	0	2
1	0	1	0	46.30	0	0
1	0	1	0	0.00	0	0
1	0	1	0	0.00	0	1
0	0	0	0	0.00	1	0
1	0	1	0	0.00	1	0
1	0	1	0	0.00	1	2

TABLE 6.4. (*Continued*)

UIR	SIBHX	DAYCARE	CIGS	CNIN	FALL	NBER
1	0	1	0	0.00	1	0
1	0	1	0	0.00	1	1
1	0	0	0	0.00	1	2
1	1	0	0	0.00	1	6
1	0	0	0	53.46	1	0
1	0	1	0	0.00	1	0
1	0	0	0	0.00	1	0
1	0	1	0	3.46	1	3
1	0	1	0	125.00	1	1
0	0	1	0	0.00	1	0
1	0	1	0	0.00	1	2
1	0	1	0	0.00	1	3
1	0	1	0	0.00	1	0
1	0	0	0	0.00	1	2
1	0	0	0	0.00	1	0
1	1	0	0	0.00	1	1
0	0	0	0	0.00	1	0
0	0	1	0	0.00	1	0
1	0	0	0	0.00	1	0
1	0	0	0	2.80	1	0
0	0	1	0	1950.00	1	0
1	0	1	0	69.44	1	2
1	0	1	0	0.00	1	3
1	0	0	0	0.00	1	0
1	0	0	0	0.00	1	2
1	1	0	0	0.00	1	0
1	1	1	0	0.00	1	4
1	0	0	0	0.00	0	0
1	0	1	0	0.00	0	0
1	0	1	0	0.00	0	1
1	0	0	0	0.00	0	1
1	0	1	0	0.00	0	1
0	0	1	0	31.53	0	0
0	0	1	0	0.00	0	0
1	0	1	0	11.40	0	3
0	0	0	0	0.00	0	0
0	1	0	0	750.00	0	0
1	0	0	0	0.00	0	0
0	0	0	0	0.00	0	1
1	1	0	0	0.00	0	0

TABLE 6.4. (*Continued*)

UIR	SIBHX	DAYCARE	CIGS	CNIN	FALL	NBER
1	1	0	0	0.00	0	0
0	0	0	0	0.00	1	0
1	1	0	0	0.00	1	2
0	0	0	0	0.00	1	1
1	0	1	0	0.00	1	1
0	1	1	22	824.22	1	1
1	0	0	0	0.00	1	0
0	0	0	0	0.00	1	2
1	1	0	0	0.00	1	1
1	1	1	25	384.98	1	2
1	0	1	0	0.00	1	2
0	0	0	0	0.00	1	0
1	0	1	0	29.41	1	0
1	0	0	0	0.00	1	0
1	0	0	0	0.00	1	0
0	0	1	0	0.00	1	0
0	0	0	0	35.59	1	0
0	0	0	0	0.00	1	0
1	0	1	0	0.00	1	3
1	0	1	0	0.00	1	4
1	0	0	0	0.00	1	1
1	0	0	0	0.00	1	0
1	0	0	0	0.00	1	1
1	1	1	35	390.80	1	2
1	0	1	0	0.00	1	0
1	0	1	0	0.00	1	2
0	0	1	0	0.00	1	0
0	0	1	0	0.00	1	0
1	1	1	0	0.00	1	3
1	1	0	22	1101.45	1	3
1	0	0	0	0.00	1	2
0	0	1	0	0.00	1	0
1	0	1	0	57.14	1	0
1	1	1	40	306.23	1	2
1	0	1	0	300.00	1	6
1	0	1	0	0.00	1	2
0	1	1	0	0.00	1	0
0	0	0	0	43.86	1	0
0	0	0	0	0.00	1	3
1	1	0	0	0.00	1	2

TABLE 6.4. (*Continued*)

UIR	SIBHX	DAYCARE	CIGS	CNIN	FALL	NBER
1	1	0	0	0.00	1	3
0	0	0	0	0.00	1	0
0	0	0	0	0.00	1	0
1	0	0	0	0.00	1	2
1	0	1	0	0.00	1	0
1	1	1	0	0.00	1	2
0	1	1	10	1000.00	1	1
0	1	0	10	0.00	1	0
0	1	1	1	0.00	0	0
1	0	0	0	0.00	0	1
1	0	0	0	0.00	0	3
1	0	0	0	0.00	0	0
0	0	0	0	0.00	0	0
0	0	1	0	0.00	0	0
1	0	0	0	0.00	0	1
0	0	0	0	0.00	0	3
1	0	0	0	0.00	0	1
0	1	1	23	400.00	0	1
1	1	1	0	0.00	0	1
0	1	0	10	0.00	0	0
1	0	1	0	0.00	0	3
0	0	1	0	0.00	0	1
1	0	1	0	0.00	0	3
0	0	1	0	0.00	0	1
1	1	1	0	0.00	0	0
0	0	0	0	0.00	0	0
0	0	0	0	0.00	0	1
0	0	1	10	1067.57	0	1
1	1	1	3	1492.31	0	0
0	0	1	0	0.00	0	2
1	0	0	0	0.00	0	0
1	0	0	0	0.00	0	0
1	0	1	0	9.41	0	1
1	0	0	0	0.00	0	0
1	0	1	0	9.84	0	2
1	0	1	10	723.58	0	2
1	0	0	0	0.00	0	2
0	0	0	0	15.63	0	0
1	0	0	0	0.00	0	0
1	1	1	30	106.60	0	0

TABLE 6.4. (*Continued*)

UIR	SIBHX	DAYCARE	CIGS	CNIN	FALL	NBER
0	0	0	0	0.00	0	0
0	0	1	0	0.00	0	0
1	0	1	0	0.00	0	1
1	0	0	0	0.00	0	0
1	0	1	0	0.00	0	0
0	0	1	0	0.00	0	0
0	0	1	0	0.00	0	0
1	1	0	0	0.00	0	0
1	0	1	0	0.00	0	1
1	0	1	0	0.00	0	1
1	0	0	0	0.00	0	0
1	0	1	0	0.00	0	2
0	1	1	30	15375.00	0	0
0	1	0	75	11000.00	0	0
0	1	1	0	0.00	0	0
1	0	1	0	0.00	0	1
1	0	0	0	0.00	0	1
0	0	0	0	0.00	0	0
0	0	1	0	17.39	0	0
0	0	0	0	0.00	0	0
0	1	1	0	0.00	0	0
0	0	0	0	0.00	0	0
1	0	0	0	0.00	0	0
0	0	1	0	0.00	0	0
0	1	1	6	44.19	0	0
1	1	0	1	0.00	0	0
0	0	1	0	0.00	0	1
1	1	1	30	447.15	0	5
0	0	0	0	0.00	0	0
0	1	0	20	230.43	0	1
1	1	1	0	0.00	0	1
0	0	1	0	0.00	0	0
0	0	1	0	0.00	0	0
0	0	1	0	217.82	0	0
0	0	1	0	0.00	0	0
1	0	0	0	0.00	0	0
1	0	1	0	32.41	0	0
1	1	0	0	0.00	0	0
1	1	1	8	43.22	0	0
1	1	1	28	664.77	0	2
1	0	1	0	0.00	0	0

CHAPTER 7

Transition from Categorical to Survival Data

Methodology discussed in this text has been directed to the analyses of categorical data; this chapter is an exception. The topics covered here, basic survival analysis and Cox's proportional hazards regression, were developed to deal with survival data; it is included here for a number of reasons. First, the border line between categorical and survival data is rather vague, especially for beginning students. Survival analysis is focused on the occurrence of an *event*, such as death—a binary outcome. Therefore, for beginners, it may be confused with the type of data that require the logistic regression analysis of Chapter 4. The basic difference is that, for survival data, studies have staggered entry and subjects are followed for varying lengths of time; they do not have the same probability for the event to occur even if they have identical characteristics, a basic assumption of the logistic regression model. Also, statistical tests for the comparison of survival distributions are special forms of the Mantel–Haenszel method of section 2.4. In addition, the coverage of Cox's proportional hazards regression (Cox, 1972) would enrich or supplement Chapter 4 because, under certain special cases, this model and the conditional logistic regression model correspond to the same likelihood function and are analyzed using the same computer program (Le and Lindgren, 1988). Finally, the inclusion of some introductory survival analysis at the end of this book is a natural extension; most methodologies used in survival analysis are generalizations of those for categorical data. For those students in applied fields, such as epidemiology, access to this methodology would be beneficial because most may not be adequately prepared for the level of sophistication of a full course in survival analysis (Le, 1997a). Sections 7.1 and 7.2 introduce some basic concepts and techniques of survival analysis; Cox's regression models are covered in sections 7.3 and 7.4.

7.1. SURVIVAL DATA

In prospective studies, the important feature is not only the outcome event, such as death, but the time to that event, the *survival time*. In order to determine the survival time T, three basic elements are needed:

1. a time origin or starting point,
2. an ending event of interest, and
3. a measurement scale for the passage of time;

for example, the life span T from birth (starting point) to death (ending event) in years (measurement scale). (See Figure 7.1.)

The time origin or starting point should be precisely defined but it need not be birth; it could be the start of a new treatment (randomization date in a clinical trial) or the admission to a hospital or a nursing home. The ending event should also be precisely defined

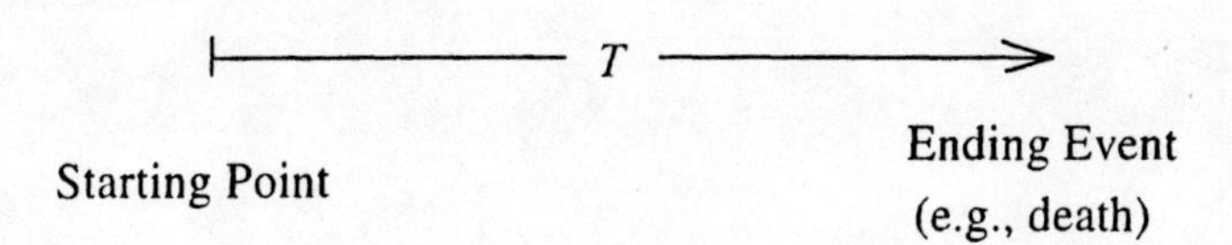

FIGURE 7.1. The Survival Time

but it need not be death; a nonfatal event such as the relapse of a disease (e.g., leukemia) or the relapse from a smoking cessation or the discharge to the community from a hospital or a nursing home satisfy the definition and are acceptable choices. The use of calendar time in health studies is common and meaningful; however, other choices for a time scale are justified—for example, hospital cost (in dollars) from admission (starting point) to discharge (ending event).

The distribution of the survival time T from enrollment or starting point to the event of interest, considered as a random variable, is characterized by either one of two equivalent functions: the *survival function* and the *hazard function*.

The survival function, denoted $S(t)$, is defined as the probability that an individual survives longer than t units of time:

$$S(t) = \Pr(T > t).$$

$S(t)$ is also known as the *survival rate*; for example, if times are in years, then $S(2)$ is the 2-year survival rate, $S(5)$ is the 5-year survival rate, and so on. The graph of $S(t)$ versus t is called the *survival curve*. as shown in Figure 7.2.

The *hazard* or *risk function* $\lambda(t)$ gives the *instantaneous* failure rate assuming that the individual has survived to time t,

$$\lambda(t) = \lim_{\delta \downarrow 0} \frac{\Pr(t \leq T \leq t + \delta \mid t \leq T)}{\delta}$$

or

$$\lambda(t)\,dt = \Pr(t \leq T \leq t + dt \mid t \leq T).$$

Thus, for a small time increment δ, the probability of an event occurring during time interval $(t, t + \delta)$ is given approximately by

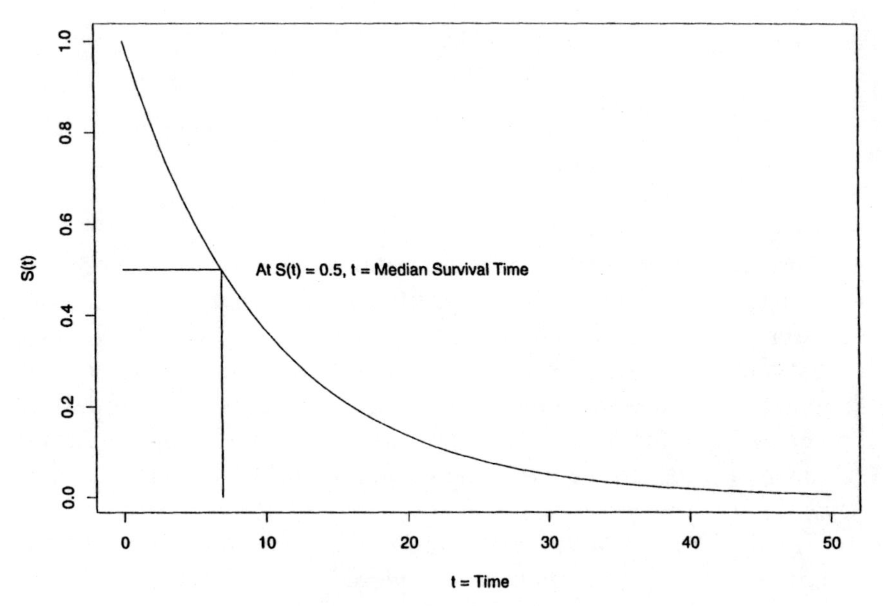

FIGURE 7.2. Survival Curve

$$\lambda(t) \cong \frac{\Pr(t \leq T \leq t + \delta \mid t \leq T)}{\delta}.$$

In other words, the hazard or risk function $\lambda(t)$ approximates the proportion of subjects dying or having events per unit time around time t. Note that this differs from the density function represented by the usual histogram; in the case of $\lambda(t)$ the numerator is a *conditional* probability. $\lambda(t)$ is also known as the *force of mortality* and is a measure of the proneness to failure as a function of age of the individual. When a population is subdivided into two subpopulations, E (exposed) and E' (nonexposed), by the presence or absence of a certain characteristic (an exposure such as smoking), each subpopulation corresponds to a hazard or risk function and the ratio of two such functions,

$$\text{RR}(t) = \frac{\lambda(t; E)}{\lambda(t; E')},$$

is called the *relative risk* of exposure to factor E. In general, the relative risk $\text{RR}(t)$ is a function of time and measures the magnitude of an effect; when it remains constant, $\text{RR}(t) = \rho$, we have a *proportional hazards model* (PHM),

$$\lambda(t;E) = \rho\lambda(t;E'),$$

with the risk of the nonexposed subpopulation served as the baseline. This is a multiplicative model; another way to express this model is

$$\lambda(t) = \lambda_0(t)e^{\beta x},$$

where $\lambda_0(t)$ is $\lambda(t;E')$—the hazard function of the unexposed subpopulation—and the indicator (or *covariate*) x is defined as

$$x = \begin{cases} 0 & \text{if unexposed} \\ 1 & \text{if exposed.} \end{cases}$$

The "regression coefficient" β represents the relative risk on the log scale. This model works with any covariate X—continuous or categorical; the above binary covariate is only a very special case. Of course, the model can be extended to include several covariates; it is usually referred to as *Cox's regression model*.

A special source of difficulty in the analysis of survival data is the possibility that some individuals may not be observed for the full time to failure or event. The so-called "random censoring" arises in medical applications with animal studies, epidemiological applications with human studies, or clinical trials. In these cases, observation is terminated before the occurrence of the event. In a clinical trial, for example, patients may enter the study at different times; then each is treated with one of several possible therapies after a randomization. We want to observe their lifetimes from enrollment, but censoring may occur in one of the following forms:

- Loss to follow-up—the patient may decide to move elsewhere.
- Dropout—a therapy may have such bad effects that it is necessary to discontinue the treatment.
- Termination of the study (for data analysis at a predetermined time).
- Death due to a cause not under investigation (for example, suicide).

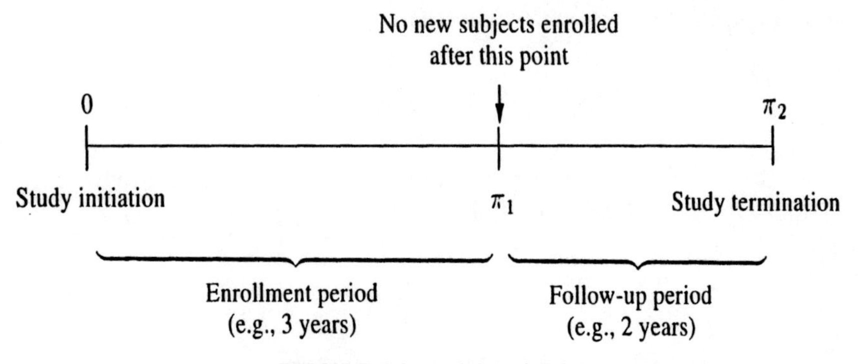

FIGURE 7.3. A Clinical Trial

Figure 7.3 shows a description of a typical clinical trial. To make it possible for statistical analysis we make the crucial assumption that, conditionally on the values of any explanatory variables (or covariates), the prognosis for any individual who has survived to a certain time t should not be affected if the individual is censored at t. That is, an individual who is censored at t should be representative of all those subjects with the same values of the explanatory variables who survive to t. In other words, survival condition and reason of loss are independent; under this assumption, there is no need to distinguish the above four forms of censoring.

We assume that observations available on the failure time of n individuals are usually taken to be independent. At the end of the study, our sample consists of n pairs of numbers (t_i, δ_i). Here δ_i is an indicator variable for survival status ($\delta_i = 0$ if the ith individual is censored; $\delta_i = 1$ if the ith individual failed) and t_i is the time to failure/event (if $\delta_i = 1$) or the censoring time (if $\delta_i = 0$); t_i is also called the *duration time*. We may also consider, in addition to t_i and δ_i, $(x_{1i}, x_{2i}, \ldots, x_{ki})$, a set of k covariates associated with the ith individual representing cofactors such as age, sex, treatment, and so on.

7.2. INTRODUCTORY SURVIVAL ANALYSIS

This section will briefly introduce a popular method for the estimation of the survival function and a family of statistical tests for the comparison of survival distributions.

7.2.1. Kaplan–Meier Curve

This section introduces the *product-limit* (PL) method of estimating the survival rates; this is also called the *Kaplan–Meier* method (Kaplan and Meier, 1958).

Let

$$t_1 < t_2 < \cdots < t_k$$

be the distinct observed death times in a sample of size n from a homogeneous population with survival function $S(t)$ to be estimated ($k \leq n$). Let n_i be the number of subjects at risk at a time just prior to t_i ($1 \leq i \leq k$; these are cases whose duration time is at least t_i), and d_i the number of deaths at t_i. The survival function $S(t)$ is estimated by

$$\hat{S}(t) = \prod_{t_i \leq t} \left(1 - \frac{d_i}{n_i}\right),$$

which is called the *product-limit estimator* or *Kaplan–Meier's estimator* with a 95 percent confidence given by

$$\hat{S}(t)\exp[\pm 1.96\hat{s}(t)],$$

where

$$\hat{s}^2(t) = \sum_{t_i \leq t} \frac{d_i}{n_i(n_i - d_i)}$$

(Greenwood, 1926).

Example 7.1. The remission times of 42 patients with acute leukemia were reported from a clinical trial undertaken to assess the ability of a drug called 6-mercaptopurine (6-MP) to maintain remission (Freireich et al., 1963; data were taken from Cox and Oakes, 1984). Each patient was randomized to receive either 6-MP or a placebo. The study was terminated after one year; patients have different follow-up times because they were enrolled sequentially at different times. Times in weeks

TABLE 7.1. Survival Rates for 6-MP Group of Example 7.1

(1)	(2)	(3)	(4)	(5)
t_i	n_i	d_i	$1 - \frac{d_i}{n_i}$	$\hat{S}(t_i)$
6	21	3	.8571	.8571
7	17	1	.9412	.8067
10	15	1	.9333	.7529
13	12	1	.9167	.6902
16	11	1	.9091	.6275
22	7	1	.8571	.5378
23	6	1	.8333	.4482

were

6-MP group: 6, 6, 6, 7, 10, 13, 16, 22, 23, 6+, 9+, 10+, 11+, 17+, 19+, 20+, 25+, 32+, 32+, 34+, 35+

Placebo group: 1, 1, 2, 2, 3, 4, 4, 5, 5, 8, 8, 8, 8, 11, 11, 12, 12, 15, 17, 22, 23

in which a t+ denotes a censored observation (i.e., the case was censored after t weeks without a relapse). For example, 10+ is a case enrolled 10 weeks before study termination and still remission-free at termination.

According to the product-limit method, survival rates for the 6-MP group are calculated by constructing a table such as Table 7.1 with five columns; to obtain $\hat{S}(t)$, multiply all values in column 4 up to and including t. From Table 7.1, we have, for example:

7-week survival rate is 80.67%

22-week survival rate is 53.78%

and a 95 percent confidence interval for $S(7)$ is (.6804, .9565).

Note: An SAS program would include these instructions:

```
PROC LIFETEST METHOD=KM;
TIME WEEKS*RELAPSE(0);
```

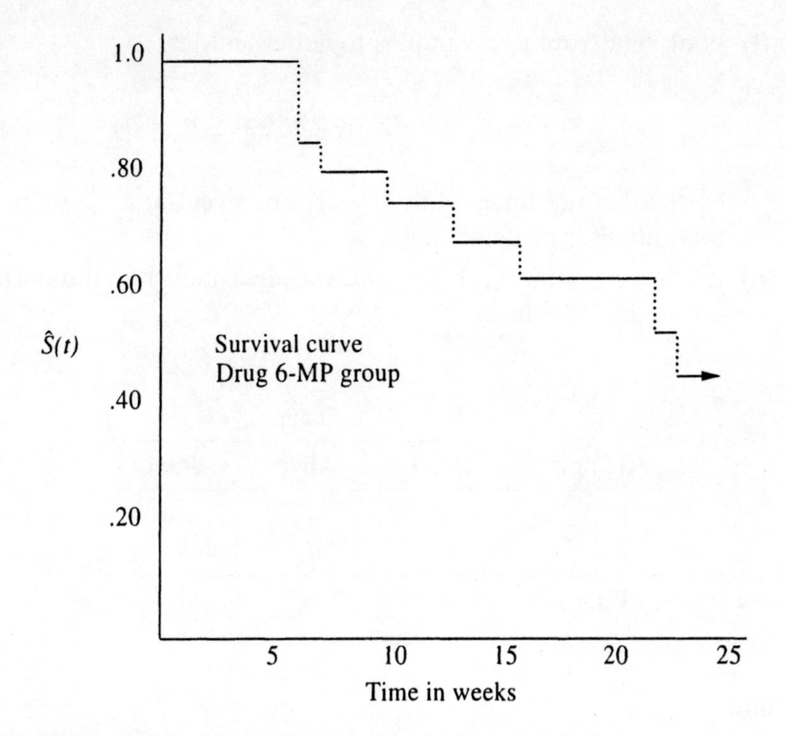

where WEEKS is the variable name for duration time, RELAPSE is the variable name for survival status, 0 is the coding for censoring, and KM stands for Kaplan–Meier method.

7.2.2. Comparison of Survival Distributions

Suppose that there are n_1 and n_2 individuals corresponding to two treatment groups 1 and 2, respectively. The study provides two samples of survival data:

$$\{(t_{1i}, \delta_{1i})\}; \qquad i = 1, 2, \ldots n_1$$

and

$$\{(t_{2j}, \delta_{2j})\}; \qquad j = 1, 2, \ldots n_2.$$

In the presence of censored observations, tests of significance can be constructed as follows:

(i) Pool data from two samples together and let

$$t_1 < t_2 < \cdots < t_m; \qquad m \leq d \leq n_1 + n_2$$

be the distinct times with at least one event at each (d is the total number of deaths).

(ii) At ordered time t_i, $1 \leq i \leq m$, the data may be summarized into a 2×2 table:

	Status		
Sample	Dead	Alive	Total
1	d_{1i}	a_{1i}	n_{1i}
2	d_{2i}	a_{2i}	n_{2i}
Total	d_i	a_i	n_i

where

n_{1i} = number of subjects from sample 1 who were at risk just before time t_i

n_{2i} = number of subjects from sample 2 who were at risk just before time t_i

$n_i = n_{1i} + n_{2i}$

d_i = number of deaths at t_i, d_{1i} of them from sample 1 and d_{2i} of them from sample 2

$= d_{1i} + d_{2i}$

$a_i = n_i - d_i$

$= a_{1i} + a_{2i}$

= number of survivors

$$d = \sum d_i.$$

In this form, the null hypothesis of equal survival functions implies the independence of "sample" and "status" in the above cross-classification 2×2 table. Therefore, under the null hypothesis, the "expected value" of d_{1i} is

$$E_0(d_{1i}) = \frac{n_{1i}d_i}{n_i}$$

(d_{1i} being the "observed value"). The variance is estimated by (hypergeometric model):

$$\mathrm{Var}_0(d_{1i}) = \frac{n_{1i}n_{2i}a_i d_i}{n_i^2(n_i - 1)}.$$

After constructing a 2×2 table for each uncensored observation, the evidence against the null hypothesis can be summarized in the following statistic,

$$\theta = \sum_{i=1}^{m} w_i[d_{1i} - E_0(d_{1i})],$$

where w_i is the "weight" associated with the 2×2 table at t_i. We have under the null hypothesis:

$$\begin{aligned} E_0(\theta) &= 0 \\ \mathrm{Var}_0(\theta) &= \sum_{i=1}^{m} w_i^2 \mathrm{Var}_0(d_{1i}) \\ &= \sum_{i=1}^{m} \frac{w_i^2 n_{1i}n_{2i}a_i d_i}{n_i^2(n_i - 1)}. \end{aligned}$$

The evidence against the null hypothesis is summarized in the standardized statistic,

$$z = \theta / \{\mathrm{Var}_0(\theta)\}^{1/2},$$

which is referred to the standard normal percentile $z_{1-\alpha}$ for a specified size α of the test. We may also refer z^2 to a chi-square distribution at one degree of freedom.

There are two important special cases:

(i) The choice

$$w_i = n_i$$

gives the generalized Wilcoxon test (also called the Gehan–Breslow test), (Gehan, 1965a; Breslow, 1970); it is reduced to the Wilcoxon test in the absence of censoring.

(ii) The choice

$$w_i = 1$$

gives the log-rank test (also called the Cox–Mantel test; it is similar to the Mantel–Haenszel procedure for the combination of several 2×2 tables in the analysis of categorical data, Mantel and Haenszel, 1959; Cox, 1972; Tarone and Ware, 1977).

There are a few other interesting issues:

1. Which test should we use? The generalized Wilcoxon statistic puts more weight on the beginning observations, and because of that its use is more powerful in detecting the effects of short-term risks. On the other hand, the log-rank statistic puts equal weight on each observation and therefore, by default, is more sensitive to exposures with a constant relative risk (proportional hazards effect; in fact, we have derived the log-rank test as a score test using the proportional hazards model). Because of these characteristics, applications of both tests may reveal not only whether an exposure has any effect, but also the nature of the effect, short-term or long-term.
2. Because of the way the tests are formulated (terms in the summation are not squared),

$$\sum_{\text{all } i} w_i[d_{1i} - E_0(d_{1i})],$$

they are only powerful when one risk is greater than the other at all times. Otherwise, some terms in this sum are positive and some terms are negative, and they cancel each other out. For example, the tests are virtually powerless for the case of crossing survival curves; in this case the assumption of proportional hazards is severely violated.

3. Some cancer treatments (bone marrow transplantation, for example) are thought to have cured patients within a short time of initiation. Then, instead of all patients having the same hazard, a biologically more appropriate model, the cure model, assumes that an unknown proportion $(1-\pi)$ are still at risk whereas the remaining proportion (π) have essentially no risk. If the aim of the study is to compare the cure proportions π's, then neither the generalized Wilcoxon nor log-rank tests are appropriate (low power). One may simply choose a time point t far enough for the curves to level off, then compare the estimated survival rates by referring to percentiles of the standard normal distribution:

$$z = \frac{\hat{S}_1(t) - \hat{S}_2(t)}{\{\mathrm{Var}[\hat{S}_1(t)] + \mathrm{Var}[\hat{S}_2(t)]\}^{1/2}}.$$

Estimated survival rates, $\hat{S}_i(t)$, and their variances are obtained as in section 7.2.1 (Kaplan–Meier procedure).

Example 7.2. Refer back to the clinical trial to evaluate the effect of 6-mercaptopurine (6-MP) to maintain remission from acute leukemia (Example 7.1). The results of the tests indicate a highly significant difference between survival patterns of the two groups. The generalized Wilcoxon test shows a slightly larger statistic indicating that the difference is slightly larger at earlier times; however, the log-rank test is almost equally significant indicating that the use of 6-MP has a long-term effect (the effect does not wear off):

Generalized Wilcoxon: $\chi^2 = 13.46$ (1 df); $p < .0001$

Log-rank: $\chi^2 = 16.79$ (1 df); $p = .0002.$

Note: An SAS program would include these instructions:

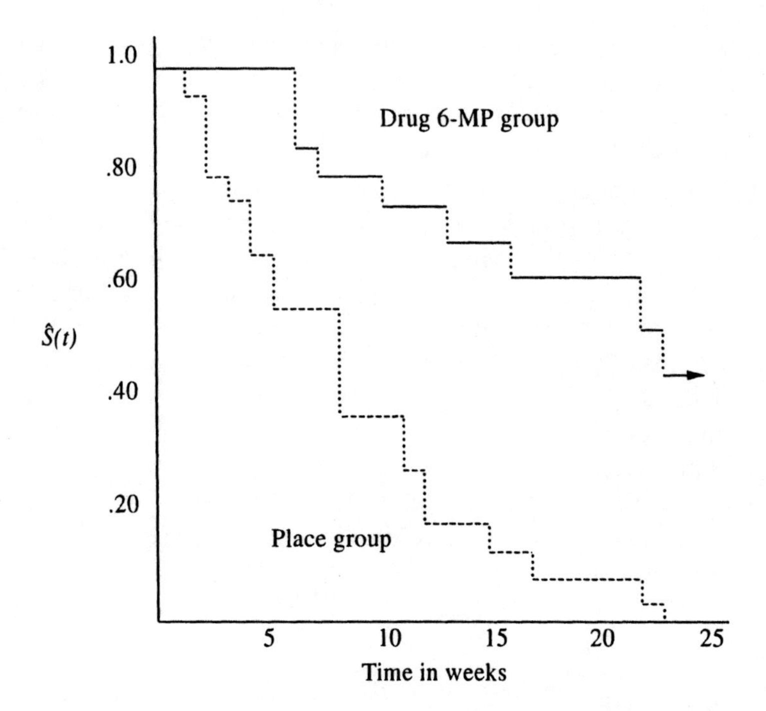


```
PROC LIFETEST METHOD=KM;
TIME WEEKS*RELAPSE(0);
STRATA DRUG;
```

where KM stands for Kaplan–Meier method, WEEKS is the variable name for duration time, RELAPSE is the variable name for survival status, 0 is the coding for censoring, and DRUG is the variable name specifying groups to be compared.

The above tests are applicable to the simultaneous comparison of several samples; when k groups are to be compared, the chi-square tests, both the log-rank and generalized Wilcoxon, have $(k-1)$ degrees of freedom.

7.3. SIMPLE REGRESSION AND CORRELATION

In this section we will discuss the basic ideas of simple regression analysis when only one predictor or independent variable is available for predicting the survival of interest; parts of this have been briefly introduced near the end of section 7.1. The following example will be used for illustration in this and the next section as well.

Example 7.3. For a group of patients who died of acute myelogenous leukemia, they were classified into the two subgroups according to the presence or absence of a morphologic characteristic of white cells. Patients termed AG positive were identified by the presence of Auer rods and/or significant granulature of the leukemic cells in the bone marrow at diagnosis. For the AG negative patients these factors were absent.

(AG Positive) $N = 17$		(AG Negative) $N = 16$	
White Blood Count (WBC)	Survival Time (weeks)	White Blood Count (WBC)	Survival Time (weeks)
2,300	65	4,400	56
750	156	3,000	65
4,300	100	4,000	17
2,600	134	1,500	7
6,000	16	9,000	16
10,500	108	5,300	22
10,000	121	10,000	3
17,000	4	19,000	4
5,400	39	27,000	2
7,000	143	28,000	3
9,400	56	31,000	8
32,000	26	26,000	4
35,000	22	21,000	3
100,000	1	79,000	30
100,000	1	100,000	4
52,000	5	100,000	43
100,000	65		

Leukemia is a cancer characterized by an overproliferation of white blood cells; the higher the white blood count (WBC), the more severe the disease. Data in the following table clearly suggest that there is such a relationship and thus, when predicting a leukemia patient's survival time it is realistic to make a prediction dependent on WBC (and any other covariates which are indicators of the progression of the disease).

7.3.1. Model and Approach

The association between two random variables X and T, the second of which—the survival time T—may be only partially observable due to right censoring, has been the focus of many investigations starting with the historical breakthrough by Cox (1972). The so-called Cox's regression model or proportional hazards model (PHM) expresses a log-linear relationship between X and the hazard function of T,

$$\begin{aligned}\lambda(t \mid X = x) &= \lim_{\delta \downarrow 0} \frac{\Pr[t \leq T \leq t + \delta \mid t \leq T,\ X = x]}{\delta} \\ &= \lambda_0(t) e^{\beta x},\end{aligned}$$

as briefly introduced near the end of section 7.1. In this model, $\lambda_0(t)$ is an *unspecified* baseline hazard (i.e., hazard at $X = 0$), and β is an *unknown* regression coefficient. The estimation of β and subsequent analyses are performed as follows. Denote the ordered distinct death times by

$$t_1 < t_2 < \cdots < t_m$$

and let R_i be the risk set just before time t_i, n_i the number of subjects in R_i, D_i the death set at time t_i, d_i the number of subjects (i.e., deaths) in D_i, and C_i the collection of all possible combinations of subjects from R_i; each combination—or subset of R_i—has d_i members, and D_i is itself one of these combinations. For example, if three subjects (A, B, and C) are at risk just before time t_i and two of them (A and B) die at t_i, then

$$\begin{aligned}R_i &= \{\text{A,B,C}\}, \qquad n_i = 3 \\ D_i &= \{\text{A,B}\}, \qquad d_i = 2 \\ C_i &= \{\{\text{A,B}\} = D_i, \{\text{A,C}\}, \{\text{B,C}\}\}.\end{aligned}$$

The number of elements in C_i is

$$\binom{n_i}{d_i}.$$

Cox (1972) suggests using

$$L = \prod_{i=1}^{m} \Pr(d_i \mid R_i, d_i)$$

$$= \prod_{i=1}^{m} \frac{\exp(\beta s_i)}{\sum_{C_i} \exp(\beta s_u)}$$

as a likelihood function, called the *partial likelihood* function, in which

$$s_i = \sum_{D_i} x_j$$

$$s_u = \sum_{D_u} x_j; \qquad D_u \in C_i.$$

An alternative likelihood, proposed by Peto (1972), is

$$L = \prod_{i=1}^{m} \frac{\sum_{D_i} \exp(\beta s_j)}{\left[\sum_{R_i} \exp(\beta x_u)\right]^{d_i}},$$

which seems to work reasonably well when the number of ties is not excessive and, therefore has become rather popular.

7.3.2. Measures of Association

We first consider the case of a binary covariate with the conventional coding

$$X_i = \begin{cases} 0 & \text{if the patient is not exposed} \\ 1 & \text{if the patient is exposed} \end{cases}.$$

Here, the term "exposed" may refer to a risk factor such as smoking, or a patient's characteristic such as race (white/nonwhite) or sex (male/female). It can be seen that, from the proportional hazards model,

$$\lambda(t;\ \text{nonexposed}) = \lambda_0(t)$$
$$\lambda(t;\ \text{exposed}) = \lambda_0(t)e^{\beta}.$$

So the ratio

$$e^{\beta} = \frac{\lambda(t;\ \text{exposed})}{\lambda(t;\ \text{nonexposed})}$$

represents the relative risk (RR) of the exposure, exposed versus nonexposed. In other words, the regression coefficient β is the value of the relative risk on the log scale.

Similarly, we have for a continuous covariate X and any value x of X,

$$\lambda(t;\ X = x) = \lambda_0(t)e^{\beta x}$$
$$\lambda(t;\ X = x + 1) = \lambda_0(t)e^{\beta(x+1)}.$$

So the ratio

$$e^{\beta} = \frac{\lambda(t;\ X = x + 1)}{\lambda(t;\ X = x)}$$

represents the relative risk (RR) due to *one unit increase* in the value of X, $X = x + 1$ versus $X = x$, for example, a systolic blood pressure of 114 mmHg versus 113 mmHg. For m units increase in the value of X, say $X = x + m$ versus $X = x$, the corresponding relative risk is $e^{m\beta}$.

The regression coefficient β can be estimated iteratively using the first and second derivatives of the partial likelihood function. From the results, we can obtained a point estimate

$$\widehat{\text{RR}} = e^{\hat{\beta}}$$

and its 95 percent confidence interval

$$\exp[\hat{\beta} \pm 1.96\text{SE}(\hat{\beta})].$$

It should be noted that the relative risk, used as a measure of association between survival time and a covariate, depends on the coding scheme for a binary factor and, for a continuous covariate X, the scale with which to measure X. For example, if we use the following coding for a factor

$$X_i = \begin{cases} -1 & \text{if the patient is not exposed} \\ 1 & \text{if the patient is exposed} \end{cases}$$

then

$$\begin{aligned} \lambda(t;\ \text{nonexposed}) &= \lambda_0(t)e^{-\beta} \\ \lambda(t;\ \text{exposed}) &= \lambda_0(t)e^{\beta}, \end{aligned}$$

so that

$$\begin{aligned} \text{RR} &= \frac{\lambda(t;\ \text{exposed})}{\lambda(t;\ \text{nonexposed})} \\ &= e^{2\beta} \end{aligned}$$

and its 95 percent confidence interval is

$$\exp[2(\hat{\beta} \pm 1.96\text{SE}(\hat{\beta}))].$$

Of course, the estimate of β under the new coding scheme is only half of that under the former scheme; therefore, the estimate of the RR remains unchanged.

The following example, however, will show the clear effect of measurement scale in the case of a continuous measurement.

Example 7.4. Refer to the data for patients with acute myelogenous leukemia in Example 7.3 and suppose we want to investigate the relationship between survival time of AG-positive patients and white blood count (WBC) in two different ways using either (i) $X = \text{WBC}$ or (ii) $X = \log(\text{WBC})$.

1. For $X = \text{WBC}$, we find

$$\hat{\beta} = .0000167,$$

from which, the relative risk for (WBC = 100,000) versus (WBC = 50,000) would be

$$\begin{aligned}\text{RR} &= \exp[(100{,}000 - 50{,}000)(.0000167)] \\ &= 2.31.\end{aligned}$$

2. For $X = \log(\text{WBC})$, we find

$$\hat{\beta} = .612331,$$

from which, the relative risk for (WBC = 100,000) versus (WBC = 50,000) would be

$$\begin{aligned}\text{RR} &= \exp\{[\log(100{,}000) - \log(50{,}000)][.612331]\} \\ &= 1.53.\end{aligned}$$

The above results are different for two different choices of X and this causes an obvious problem of choosing an appropriate measurement scale. Of course, we assume a *linear* model and one choice of X would fit better than the other (there are methods for checking this assumption).

Note: An SAS program would include these instructions:

```
PROC PHREG DATA=CANCER;
MODEL WEEKS*DEATH(0)=WBC;
```

where CANCER is the name assigned to the data set, WEEKS is the variable name for duration time, DEATH is the variable name for survival status, and 0 is the coding for censoring.

7.3.3. Tests of Association

The null hypothesis to be considered is:

$$\mathcal{H}_0 : \beta = 0.$$

The reason for interest in testing whether or not $\beta = 0$ is that $\beta = 0$ implies there is no relation between survival time T and the covariate X under investigation. For the case of a categorical covariate, the test based on the score statistic of Cox's regression model is identical to the log-rank test of section 7.2.2.

7.4. MULTIPLE REGRESSION AND CORRELATION

The effect of some factor on survival time may be influenced by the presence of other factors through effect modifications (i.e., interactions). Therefore, in order to provide a more comprehensive prediction of the future of patients with respect to duration, course, and outcome of a disease, it is very desirable to consider a large number of factors and sort out which ones are most closely related to diagnosis. In this section, we will discuss a multivariate method for risk determination. This method, which is multiple regression analysis, involves a linear combination of the explanatory or independent variables; the variables must be quantitative with particular numerical values for each patient. Information concerning possible factors is usually obtained as a subsidiary aspect from clinical trials that were designed to compare treatments. A covariate or prognostic patient characteristic may be dichotomous, polytomous, or continuous (categorical factors will be represented by dummy variables). Examples of dichotomous covariates are sex, and presence or absence of a certain co-morbidity. Polytomous covariates include race, and different grades of symptoms; these can be covered by the use of dummy variables. Continous covariates include patient age, blood pressure, and so on. In many cases, data transformations (e.g., taking the logarithm) may be desirable.

7.4.1. Proportional Hazards Models with Several Covariates

Suppose we want to consider k covariates simultaneously; the proportional hazards model of the previous section can be easily generalized and expressed as

$$
\begin{aligned}
\lambda[t \mid \mathbf{X} = (x_1, x_2, \ldots, x_k)] &= \lim_{\delta \downarrow 0} \frac{\Pr[t \leq T \leq t + \delta \mid t \leq T,\ \mathbf{X} = x]}{\delta} \\
&= \lambda_0(t) e^{[\beta^T \mathbf{x}]} \\
&= \lambda_0(t) e^{[\beta_1 x_1 + \beta_2 x_2 + \cdots + \beta_k x_k]},
\end{aligned}
$$

where $\lambda_0(t)$ is an *unspecified* baseline hazard (i.e., hazard at $X = 0$), and $\beta^T = (\beta_1, \beta_2, \ldots, \beta_k)$ are k *unknown* regression coefficients. In order to have a meaningful baseline hazard, it may be necessary to standardize continuous covariates about their means

$$\text{new } x = x - \overline{x},$$

so that $\lambda_0(t)$ is the hazard function associated with *a typical patient* (i.e., a hypothetical one having all covariates at their average values).

The estimation of β and subsequent analyses are performed similar to the univariate case using Cox's partial likelihood function

$$\begin{aligned} L &= \prod_{i=1}^{m} \Pr(d_i \mid R_i, d_i) \\ &= \prod_{i=1}^{m} \frac{\exp\left[\sum_{j=1}^{k} \beta_j s_{ji}\right]}{\sum_{C_i} \exp\left[\sum_{j=1}^{k} \beta_j s_{ju}\right]}, \end{aligned}$$

where

$$s_{ji} = \sum_{l \in D_i} x_{jl}$$

$$s_{ju} = \sum_{l \in D_u} x_{jl}; \qquad D_u \in C_i.$$

Also similar to the univariate case, $\exp(\beta_i)$ represents:

(i) the relative risk associated with an exposure if X_i is binary (exposed $X_i = 1$ versus unexposed $X_i = 0$), or
(ii) the relative risk due to one unit increase if X_i is continuous ($X_i = x + 1$ versus $X_i = x$),

and after $\hat{\beta}_i$ and its standard error have been obtained, a 95 percent confidence interval for the above relative risk is given by:

$$\exp[\hat{\beta}_i \pm 1.96\text{SE}(\hat{\beta}_i)].$$

These results are necessary in the effort to identify important prognostic or risk factors. Of course, before such analyses are done, the problem and the data have to be examined carefully. If some of the variables are highly correlated, then one or fewer of the correlated factors are likely to be as good predictors as all of them; information from other similar studies also has to be incorporated so as to drop some of these correlated explanatory variables. The uses of products, such as X_1X_2, and higher-power terms, such as X_1^2, may be necessary and can improve the goodness-of-fit. It is important to note that we are assuming a *linear* regression model in which, for example, the relative risk due to one unit increase in the value of a continuous X_i ($X_i = x + 1$ vs. $X_i = x$) is independent of x. Therefore, if this *linearity* seems to be violated, the incorporation of powers of X_i should be seriously considered. The use of products will help in the investigation of possible effect modifications. And, finally, there is the messy problem of missing data; most packaged programs would delete the patient if one or more covariate values are missing.

7.4.2. Testing Hypotheses in Multiple Regression

Once we have fitted a multiple proportional hazards regression model and obtained estimates for the various parameters of interest, we want to answer questions about the contributions of various factors to the prediction of the future of patients. There are three types of such questions:

- **(i)** An overall test: Taken collectively, does the entire set of explanatory or independent variables contribute significantly to the prediction of survivorship?
- **(ii)** Test for the value of a single factor: Does the addition of one particular factor of interest add significantly to the prediction of survivorship over and above that achieved by other factors?
- **(iii)** Test for contribution of a group of variables: Does the addition of a group of factors add significantly to the prediction of survivorship over and above that achieved by other factors?

Overall Regression Tests

We now consider the first question stated above concerning an overall test for a model containing k factors, say,

$$\begin{aligned}\lambda[t \mid \mathbf{X} = (x_1, x_2, \ldots, x_k)] &= \lim_{\delta \downarrow 0} \frac{\Pr[t \le T \le t + \delta \mid t \le T,\ \mathbf{X} = x]}{\delta} \\ &= \lambda_0(t) e^{[\beta^T \mathbf{x}]} \\ &= \lambda_0(t) e^{[\beta_1 x_1 + \beta_2 x_2 + \cdots + \beta_k x_k]}.\end{aligned}$$

The null hypothesis for this test may stated as: "All k independent variables *considered together* do not explain the variation in survival times." In other words,

$$\mathcal{H}_0 : \beta_1 = \beta_2 = \cdots = \beta_k = 0.$$

Three likelihood-based statistics can be used to test this *global* null hypothesis; each has an asymptotic chi-squared distribution with k degrees of freedom under $\mathcal{H}_0$.

(i) Likelihood test:

$$\chi^2_{\mathrm{LR}} = 2[\ln L(\hat{\beta}) - \ln L(\mathbf{0})].$$

(ii) Wald's test:

$$\chi^2_W = \hat{\beta}^T [\hat{\mathbf{V}}(\hat{\beta})]^{-1} \hat{\beta}.$$

(iii) Score test:

$$\chi^2_S = \left[\frac{\delta \ln L(\mathbf{0})}{\delta \beta}\right] \left[-\frac{\delta^2 \ln L(\mathbf{0})}{\delta \beta^2}\right]^{-1} \left[\frac{\delta \ln L(\mathbf{0})}{\delta \beta}\right].$$

All three statistics are provided by most standard computer programs.

Tests for a Single Variable

Let us assume that we now wish to test whether the addition of one particular factor of interest adds significantly to the prediction of survivorship over and above that achieved by other factors already present in the model. The null hypothesis for this test may stated as: "Factor X_i does not have any value added to the prediction of survivorship *given that other factors are already included in the model.*" In other words,

$$\mathcal{H}_0 : \beta_i = 0.$$

To test such a null hypothesis, one can perform a likelihood ratio chi-squared test, with 1 df, similar to that for the above global hypothesis:

$$\chi^2_{\text{LR}} = 2[\ln L(\hat{\beta};\ \text{all } X\text{'s}) - \ln L(\hat{\beta};\ \text{all other } X\text{'s with } X_i \text{ deleted})].$$

A much easier alternative method is using

$$z_i = \frac{\hat{\beta}_i}{\text{SE}(\hat{\beta}_i)},$$

where $\hat{\beta}_i$ is the corresponding estimated regression coefficient and $\text{SE}(\hat{\beta}_i)$ is the estimate of the standard error of $\hat{\beta}_i$, both of which are printed by standard packaged programs. In performing this test, we refer the value of the z statistic to percentiles of the standard normal distribution. This is equivalent to Wald's chi-squared test as applied to one parameter.

Contribution of a Group of Variables

This testing procedure addresses the more general problem of assessing the additional contribution of two or more factors to the prediction of survivorship over and above that made by other variables already in the regression model. In other words, the null hypothesis is of the form

$$\mathcal{H}_0 : \beta_1 = \beta_2 = \cdots = \beta_m = 0.$$

To test such a null hypothesis, one can perform a likelihood ratio chi-squared test, with m df,

$$\chi^2_{\text{LR}} = 2[\ln L(\hat{\beta};\ \text{all } X\text{'s}) - \ln L(\hat{\beta};\ \text{all other } X\text{'s with } X\text{'s under investigation deleted})].$$

As with the above z *test*, this *multiple contribution* procedure is very useful for assessing the importance of potential explanatory variables. In particular, it is often used to test whether a similar group of variables, such as *demographic characteristics*, is important for the prediction of survivorship; these variables have some trait in common. Another application would be a collection of powers *and/or* product terms (referred to as interaction variables). It is often of interest to assess the interaction effects collectively before trying to consider individual interaction terms in a model. In fact, such use reduces the total number of tests to be performed and this, in turn, helps to provide better control of overall Type I error rates, which may be inflated due to multiple testing.

Stepwise Regression

In applications, our major interest is to identify important prognostic factors. In other words, we wish to identify from many available factors a small subset of factors that relate significantly to the length of survival time of patients. In that identification process, of course, we wish to avoid a Type I (false positive) error. In a regression analysis, a Type I error corresponds to including a predictor that has no real relationship to survivorship; such an inclusion can greatly confuse the interpretation of the regression results. In a standard multiple regression analysis, this goal can be achieved by using a strategy that adds into or removes from a regression model one factor at a time according to a certain order of relative importance. The details of this stepwise process for survival data are similar to those for logistic regression in Chapter 4.

Stratification

The proportional hazards model requires that of a covariate X the hazard functions at different levels, e.g., $\lambda(t;\text{ exposed})$ and $\lambda(t;\text{ non-exposed})$, are proportional. Of course, sometimes there are factors, the different levels of which produce hazard functions which deviate markedly from proportionality. These factors may not be under investigation themselves, especially those of no intrinsic interest, those with a large number of levels, and/or those where interventions are not possible. But these factors may act as important confounders, which must be included in any meaningful analysis so as to improve predictions concerning other covariates. Common examples include sex, age, neighborhhood, and so on. To accommodate such confounders, an extension of the proportional hazards model is desirable. Suppose there is a factor that occurs on q levels and for which the proportional hazards model may be violated. If this factor is under investigation as a *covariate*, then the model and subsequent analyses are not applicable. However, if this factor is *not* under investigation and is considered only as a confounder so as to improve analyses and/or predictions concerning other covariates, then *we can treat it as a stratification factor*. By doing that we will get no results concerning this factor (which are not wanted), but in return we do not have to assume that the hazard functions corresponding to different levels are proportional (which may be severely violated). Suppose the stratification factor Z has q levels; this factor is not clinically important itself but adjustments are still needed in efforts to investigate other covariates. We define the hazard function for an individual in the jth stratum (or level) of this factor as

$$\begin{aligned}\lambda[t \mid \mathbf{X} = (x_1, x_2, \ldots, x_k)] &= \lambda_{0j}(t)e^{[\beta^T \mathbf{x}]} \\ &= \lambda_{0j}(t)e^{[\beta_1 x_1 + \beta_2 x_2 + \cdots + \beta_k x_k]}\end{aligned}$$

for $j = 1, 2, \ldots, q$ where $\lambda_0(t)$ is an *unspecified* baseline hazard for the jth stratum and $\mathbf{X}$ represents other k covariates under investigation (excluding the stratification itself). The baseline hazard functions are allowed to be arbitrary and are completely unrelated (and, of course, *not* proportional). The basic additional assumption here, which is the same as that in the *analysis of covariance*, requires that the $\beta's$ are the same across strata (i.e., the so-called "parallel lines" assumption, which is testable).

In the analysis, we identify distinct times of events for the jth stratum and form the partial likelihood $L_j(\beta)$ as in the previous sections. The *overall* partial likelihood of β is then the product of those q stratum-specific likelihoods:

$$L(\beta) = \prod_{j=1}^{q} L_j(\beta).$$

Subsequent analyses, finding maximum likelihood estimates as well as using score statistics, are straightforward. For example, if the null

TABLE 7.2. Investigation of Effect Modification

Factor	Coefficient	St. Error	z Statistic	p-Value
WBC	0.14654	0.17869	0.821	0.4122
AG-group	−5.85637	2.75029	−2.129	0.0332
Product	0.49527	0.27648	1.791	0.0732

hypothesis $\beta = 0$ for a given covariate is of interest, the score approach would produce a *stratified* log-rank test. An important application of stratification, the analysis of epidemiologic matched studies resulting in the conditional logistic regression model, was presented in Chapter 5.

Example 7.5. Refer to the myelogenous leukemia data of Example 7.3. Patients were classified into the two groups according to the presence or absence of a morphologic characteristic of white cells and the primary covariate is white blood count (WBC). Using

$$X_1 = \ln(\text{WBC})$$

$$X_2 = \text{AG-group; 0 if negative and 1 if positive}$$

we fit the following model with one interaction term:

$$\lambda[t \mid \mathbf{X} = (x_1, x_2)] = \lambda_0(t)e^{[\beta_1 x_1 + \beta_2 x_2 + \beta_3 x_1 x_2]}.$$

From the results in Table 7.2, it can be seen that the interaction effect is almost significant at the 5 percent level ($p = .0732$); i.e., the presence of the morphologic characteristic modifies substantially the effect of WBC.

7.4.3. Time-Dependent Covariates and Applications

In prospective studies, since subjects are followed over time, values of many independent variables or covariates may be changing—covariates such as patient age, blood pressure, even treatment. In general, covariates are divided into two categories: *fixed* and *time-dependent*. A covariate is time-dependent if the *difference* between covariate values from two different subjects may be changing with time. For example, Sex and Age are fixed covariates; a patient's age is increasing by one a year but the difference in age between two patients remains unchanged. On the other hand, Blood Pressure is an obvious time-dependent covariate. The following are three important groups of time-dependent covariates.

Examples

(i) Personal characteristics whose measurements are periodically made during the course of a study. Blood Pressure fluctuates; so do Cholesterol level and Weight. Smoking and Alcohol consumption habits may change.

(ii) Cumulative exposure: In many studies, exposures such as Smoking are often dichotomized; subjects are classified as exposed or unexposed. But this may be oversimplified, leading to loss of information; the length of exposure may be important. As time goes on, a nonsmoker remains a nonsmoker but "years of smoking" for a smoker increases.

(iii) Another important group is "switching treatments." In a clinical trial, a patient may be transferred from one treatment to another due to side effects or even by the patient's request. Organ transplants form another category with switching treatments; when a suitable donor is found, a subject is switched from nontransplanted group to transplanted group. The case of intensive care units is even more complicated, where a patient may be moving in and out more than once.

Implementation

Recall that in the analysis using the proportional hazards model, we order the death times and form the partial likelihood function

$$L = \prod_{i=1}^{m} \Pr(d_i \mid R_i, d_i)$$

$$= \prod_{i=1}^{m} \frac{\exp\left[\sum_{j=1}^{k} \beta_j s_{ji}\right]}{\sum_{C_i} \exp\left[\sum_{j=1}^{k} \beta_j s_{ju}\right]},$$

where

$$s_{ji} = \sum_{l \in D_i} x_{jl}$$

$$s_{ju} = \sum_{l \in D_u} x_{jl}; \qquad D_u \in C_i,$$

where R_i is the risk set just before time t_i, n_i the number of subjects in R_i, D_i the death set at time t_i, d_i the number of subjects (i.e., deaths) in D_i, and C_i the collection of all possible combinations of subjects from R_i. In this approach, we try to "explain" why *event(s)* occurred to *subject(s)* in D_i while all subjects in R_i are equally at risk; *this explanation, through the use of* s_{ji} *and* s_{ju}, *is based on the covariate values measured at time* t_i. Therefore, this needs some modification in the presence of time-dependent covariates because events at time t_i should be explained by *values of covariates measured at that particular moment*. Blood pressure, for example, measured years before may become irrelevant.

First, notations are expanded to handle time-dependent covariates. Let x_{jil} be the value of factor x_j measured from individual l at time t_i; then the above likelihood function becomes

$$L = \prod_{i=1}^{m} \Pr(d_i \mid R_i, d_i)$$

$$= \prod_{i=1}^{m} \frac{\exp\left[\sum_{j=1}^{k} \beta_j s_{jii}\right]}{\sum_{C_i} \exp\left[\sum_{j=1}^{k} \beta_j s_{jiu}\right]},$$

where

$$s_{jii} = \sum_{l \in D_i} x_{jil}$$

$$s_{jiu} = \sum_{l \in D_u} x_{jil}; \qquad D_u \in C_i.$$

From this new likelihood function, applications of subsequent steps (estimation of β's, formation of test statistics, and the estimation of the baseline survival function) are straightforward. In practical implementation, most standard computer programs have somewhat different procedures for two categories of time-dependent covariates: those that can be defined by a mathematical equation (external) and those measured directly from patients (internal); the former categories are much more easily implemented.

A Simple Test of Goodness-of-Fit

Treatment of time-dependent covariates leads to a simple test of goodness-of-fit. Consider the case of a fixed covariate, denoted by X_1. Instead of the basic proportional hazards model

$$\begin{aligned}\lambda(t \mid X_1 = x_1) &= \lim_{\delta \downarrow 0} \frac{\Pr[t \le T \le t + \delta \mid t \le T,\ X_1 = x_1]}{\delta} \\ &= \lambda_0(t) e^{\beta x_1},\end{aligned}$$

we can define an additional time-dependent covariate X_2,

$$X_2 = X_1 t,$$

consider the expanded model,

$$\begin{aligned}\lambda(t;\ X_1 = x_1) &= \lambda_0(t) e^{\beta_1 x_1 + \beta_2 x_2} \\ &= \lambda_0(t) e^{\beta_1 x_1 + \beta_2 x_1 t},\end{aligned}$$

and examine the significance of

$$\mathcal{H} : \beta_2 = 0.$$

The reason for interest in testing whether or not $\beta_2 = 0$ is that $\beta_2 = 0$ implies a goodness-of-fit of the proportional hazards model for the factor under investigation, X_1. Of course, in defining the new covariate X_2, t could be replaced by any function of t; a commonly used one is:

$$X_2 = X_1 \log(t).$$

This simple approach results in a test of a specific alternative to the proportionality. The computational implementation here is very similar to the case of cumulative exposures; however, X_1 may be binary or continuous. We may even investigate the goodness-of-fit fot several variables simultaneously.

Example 7.6. Refer to the data set in Example 7.1 where the remission times of 42 patients with acute leukemia were reported from a clinical trial undertaken to assess the ability of a drug called 6-mercaptopurine (6-MP) to maintain remission. Each patient was randomized to receive either 6-MP or a placebo. The study was terminated after one year; patients have different follow-up times because they were enrolled sequentially at different times. Times in weeks were

6-MP group: 6, 6, 6, 7, 10, 13, 16, 22, 23, 6+, 9+, 10+, 11+, 17+, 19+, 20+, 25+, 32+, 32+, 34+, 35+

Placebo group: 1, 1, 2, 2, 3, 4, 4, 5, 5, 8, 8, 8, 8, 11, 11, 12, 12, 15, 17, 22, 23

in which a $t+$ denotes a censored observation (i.e., the case was censored after t weeks without a relapse). For example, 10+ is a case enrolled 10 weeks before study termination and still remission-free at termination.

Since the proportional hazards model is often assumed in the comparison of two survival distributions such as in this example (also see Example 7.2), it is desirable to check it for validity (if the proportionality is rejected, it would lend support to the conclusion that this drug does have some cumulative effects).

Let X_1 be the indicator variable defined by

$$X_1 = \begin{cases} 0 & \text{if placebo} \\ 1 & \text{if treated by 6-MP,} \end{cases}$$

and

$$X_2 = X_1 t$$

TABLE 7.3. Investigation of Goodness-of-Fit

Factor	Coefficient	St. Error	z Statistic	p-Value
X_1	−1.55395	0.81078	−1.917	0.0553
X_2	−0.00747	0.06933	−0.108	0.9142

represent "treatment weeks" (time t is recorded in weeks). In order to judge the validity of the proportional hazards model with respect to X_1, it is the effect of this newly defined covariate X_2 that we want to investigate.

We fit the following model

$$\lambda[t \mid \mathbf{X} = (x_1, x_2)] = \lambda_0(t)e^{[\beta_1 x_1 + \beta_2 x_2]},$$

and from the results in Table 7.3, it can be seen that the "accumulation effect" or "lack of fit" represented by X_2 is insignificant; in other words, there is not enough evidence to be concerned about the validity of the proportional hazards model.

Note: An SAS program would include these instructions:

```
PROC PHREG DATA = CANCER;
MODEL WEEKS*RELAPSE(0) = DRUG TESTEE;
TESTEE = DRUG*WEEKS;
```

where WEEKS is the variable name for duration time, RELAPSE is the variable name for survival status, 0 is the coding for censoring, DRUG is the 0/1 group indicator (i.e., X_1), and TESTEE is the newly created variable (i.e., X_2).

7.5. EXERCISES

Pneumocystic Carinii Pneumonia (PCP) is the most common opportunistic infection and a life-threatening disease in HIV-infected patients. Many North Americans with AIDS have one or two episodes of PCP during the course of their HIV infection. PCP is a consideration factor in mortality, morbidity, and expense; and recurrences are common. In the following data set, we have (Table 7.4):

- Treatments, coded as A and B

TABLE 7.4. AIDS Data

TRT	CD4	SEX	RACE	WT	HOMO	PCP	PDATE	DIE	DDATE
B	2	1	1	142	1	1	11.9	0	14.6
B	139	1	2	117	0	0	11.6	1	11.6
A	68	1	2	149	0	0	12.8	0	12.8
A	12	1	1	160	1	0	7.3	1	7.3
B	36	1	2	157	0	1	4.5	0	8.5
B	77	1	1	12	1	0	18.1	1	18.1
A	56	1	1	158	0	0	14.7	1	14.7
B	208	1	2	157	1	0	24.0	1	24.0
A	40	1	1	122	1	0	16.2	0	16.2
A	53	1	2	125	1	0	26.6	1	26.6
A	28	1	2	130	0	1	14.5	1	19.3
A	162	1	1	124	0	0	25.8	1	25.8
B	163	1	2	130	0	0	16.1	0	16.1
A	65	1	1	12	0	0	19.4	0	19.4
A	247	1	1	167	0	0	23.4	0	23.4
B	131	1	1	16	0	0	2.7	0	2.7
A	25	1	1	13	1	0	20.1	1	20.1
A	118	1	1	155	0	0	17.3	0	17.3
B	21	1	1	126	0	0	6.0	1	6.0
B	81	1	2	168	1	0	1.6	0	1.6
A	89	1	1	169	1	0	29.5	0	29.5
B	172	1	1	163	1	0	24.2	1	16.5
B	21	1	1	164	1	1	4.9	0	22.9
A	7	1	1	139	1	0	14.8	1	14.8
B	94	1	1	165	1	0	29.8	0	29.8
B	14	1	2	17	0	0	21.9	1	21.4
A	38	1	1	17	1	0	18.8	1	18.8
A	73	1	1	14	1	0	20.5	1	20.5
B	25	1	2	19	0	0	13.1	1	13.1
A	13	1	3	121	1	0	21.4	0	21.4
B	30	1	1	145	1	0	21.0	0	21.0
A	152	1	3	124	1	0	19.4	0	19.4
B	68	1	3	15	0	0	17.5	1	17.4
A	27	1	1	128	1	0	18.5	0	18.5
A	38	1	1	159	1	1	18.3	1	18.3
B	265	1	3	242	0	0	11.1	1	11.1
B	29	0	3	13	0	0	14.0	0	14.0
A	73	1	3	130	1	0	11.1	1	11.1
B	103	1	1	164	1	0	11.1	0	11.1
B	98	1	1	193	1	0	10.2	0	10.2
A	120	1	1	17	1	0	5.2	0	5.2
B	131	1	2	184	0	0	5.5	0	5.5
A	48	0	1	160	0	0	13.7	1	13.7
B	80	1	1	115	1	0	12.0	0	12.0

TABLE 7.4. (*Continued*)

TRT	CD4	SEX	RACE	WT	HOMO	PCP	PDATE	DIE	DDATE
A	132	1	3	130	1	0	27.0	0	27.0
A	54	1	1	148	1	0	11.7	0	11.7
B	189	1	1	198	1	0	24.5	0	24.5
B	14	1	2	160	0	0	1.3	0	1.3
A	321	1	1	130	1	1	18.5	0	18.6
B	148	1	1	126	1	0	22.8	0	22.8
B	54	1	1	181	1	1	14.5	0	15.7
A	17	1	1	152	1	0	19.8	0	19.8
A	37	1	3	120	1	0	16.6	0	16.6
B	71	0	1	136	0	0	16.5	0	16.5
A	9	1	3	130	1	0	8.9	1	8.9
B	231	1	3	140	0	1	10.3	0	10.6
A	22	1	2	190	1	0	8.5	0	8.5
A	43	1	1	134	1	0	17.9	1	17.9
B	103	1	1	110	1	0	20.3	0	20.3
A	146	1	1	213	1	0	20.5	0	20.5
A	92	1	1	128	1	0	14.2	0	14.2
B	218	1	1	163	1	0	1.9	0	1.9
A	100	1	1	170	1	0	14.0	0	14.0
B	148	1	1	158	1	1	15.4	0	16.1
B	44	1	2	124	1	0	7.3	0	7.3
A	76	1	1	149	0	1	15.9	1	23.4
B	30	1	1	181	1	0	6.6	1	6.6
B	260	1	1	165	1	1	7.5	1	18.0
B	40	1	1	204	0	0	21.0	1	18.8
A	90	1	1	149	1	1	17.0	0	21.8
A	120	1	1	152	0	0	21.8	0	21.8
B	80	1	1	199	1	1	20.6	0	20.6
A	170	1	1	141	1	0	18.6	1	18.6
A	54	1	1	148	1	0	18.6	0	18.6
A	151	1	1	140	1	0	21.2	0	21.2
B	107	1	1	158	1	0	22.5	1	22.3
A	9	1	1	116	1	0	18.0	1	18.0
B	79	1	3	132	0	0	22.6	1	22.6
A	72	1	1	131	1	0	19.9	0	19.9
A	100	1	1	182	1	0	21.2	0	21.2
B	16	1	2	106	1	0	18.3	0	18.3
B	10	1	1	168	1	0	24.7	0	24.7
A	135	1	1	149	1	0	23.8	0	23.8
B	235	1	1	137	0	0	22.7	0	22.7
B	20	1	1	104	0	0	14.0	0	14.0
A	67	1	2	150	0	0	19.4	1	19.4
B	7	1	1	182	1	0	17.0	1	17.0
B	139	1	1	143	1	0	21.4	0	21.4

TABLE 7.4. (*Continued*)

TRT	CD4	SEX	RACE	WT	HOMO	PCP	PDATE	DIE	DDATE
B	13	1	3	132	0	0	23.5	0	23.5
A	117	1	1	144	1	0	19.5	1	19.5
A	11	1	2	111	1	0	19.3	1	19.3
B	280	1	1	145	1	0	11.6	1	11.6
A	119	1	1	159	1	1	13.1	0	19.3
B	9	1	1	146	1	1	17.0	1	18.2
A	30	1	2	150	0	0	20.9	0	20.9
B	22	1	1	138	1	1	1.1	1	10.0
B	186	1	3	114	1	0	17.2	0	17.2
A	42	1	1	167	1	0	19.2	0	19.2
B	9	1	2	146	1	0	6.0	1	6.0
B	99	1	1	149	0	1	14.8	0	15.5
A	21	1	1	141	1	0	17.7	0	17.7
A	16	1	2	116	0	0	17.5	0	17.5
B	10	1	1	143	1	1	8.3	1	13.5
B	109	1	1	130	1	1	12.0	0	12.1
B	72	1	1	137	0	0	12.8	0	12.8
B	582	1	1	143	1	0	15.7	0	15.7
A	8	1	2	134	1	0	9.3	1	9.3
A	69	1	1	160	0	0	10.1	0	10.1
A	57	1	1	138	1	0	10.2	0	10.2
A	47	1	1	159	1	0	9.1	0	9.1
A	149	1	3	152	0	0	9.8	0	9.8
B	229	1	2	130	1	0	9.4	0	9.4
A	9	1	1	165	1	0	9.2	0	9.2
A	10	1	1	162	0	0	9.2	0	9.2
A	78	1	1	145	1	0	10.2	0	10.2
B	147	1	1	180	1	0	9.0	0	9.0
B	147	1	1	180	1	0	9.0	0	9.0
B	126	1	1	124	1	0	5.5	0	5.5
A	19	1	2	192	0	0	6.0	0	6.0
A	142	1	1	17	1	0	17.3	1	17.2
B	277	0	1	14	0	0	17.0	0	17.0
B	80	1	1	13	1	0	15.0	0	15.0
A	366	1	1	15	1	0	14.9	0	14.9
A	76	1	1	18	1	0	9.2	0	9.2
A	13	1	1	171	1	0	30.2	0	30.2
B	17	1	1	276	0	0	15.8	1	15.8
B	193	1	1	164	1	0	22.5	1	22.5
A	108	1	1	161	0	0	24.0	0	24.0
B	41	1	1	153	0	0	23.9	0	23.9
A	113	1	1	131	0	0	21.4	0	21.4
B	1	1	2	136	0	0	19.6	0	19.6
A	.	0	2	109	0	0	6.0	1	6.0

TABLE 7.4. (*Continued*)

TRT	CD4	SEX	RACE	WT	HOMO	PCP	PDATE	DIE	DDATE
A	47	1	1	168	1	0	18.2	1	18.2
B	172	1	2	195	1	0	10.3	1	10.3
A	247	1	1	123	1	0	16.2	0	16.2
B	21	1	2	124	0	0	9.7	0	9.7
B	0	1	2	116	1	0	11.0	0	11.0
B	38	1	2	160	1	0	14.7	0	14.7
A	50	1	1	127	1	0	13.6	0	13.6
A	4	1	2	218	0	0	12.9	0	12.9
A	150	1	1	200	1	0	11.7	1	11.7
B	0	1	2	133	0	1	4.7	0	11.8
A	97	1	2	156	0	0	11.9	0	11.9
B	312	1	1	140	1	0	10.6	0	10.6
B	35	1	1	155	1	0	11.0	0	11.0
A	100	1	1	157	1	0	9.2	0	9.2
A	69	1	1	126	0	0	9.2	0	9.2
B	124	1	2	135	1	0	6.5	0	6.5
B	25	1	1	162	1	0	16.0	1	0.0
A	61	0	2	102	0	0	18.5	1	18.5
B	102	1	1	177	1	1	11.3	0	17.4
A	198	1	2	164	0	0	23.2	0	23.2
B	10	1	1	173	0	0	8.4	1	8.4
A	56	1	1	163	1	0	11.9	0	11.9
A	43	1	1	134	1	0	9.2	0	9.2
B	202	1	2	158	0	0	9.2	0	9.2
A	31	1	1	150	1	0	9.5	1	5.6
B	243	1	1	136	1	0	22.7	0	22.7
B	40	1	1	179	1	0	23.0	0	23.0
A	365	1	1	129	0	0	17.9	0	17.9
A	29	1	2	145	0	1	0.6	0	2.6
A	97	1	1	127	0	0	13.7	0	13.7
B	314	1	3	143	0	0	12.2	1	12.2
B	17	1	1	114	0	1	17.3	0	17.7
A	123	1	1	158	0	0	21.5	0	21.5
B	92	1	1	128	0	0	6.0	0	6.0
A	39	0	2	15	0	0	10.8	0	10.8
A	87	1	1	156	1	0	28.3	0	28.3
A	93	1	1	170	0	0	23.9	0	23.9
A	4	0	2	104	0	0	21.0	0	21.0
A	60	1	1	150	0	0	6.3	0	6.3
B	20	0	1	133	0	0	17.3	1	17.3
A	52	1	3	125	0	0	12.0	0	12.0
B	78	0	1	99	0	0	16.7	0	16.7
B	262	1	2	192	0	0	12.7	0	12.7
A	19	1	2	143	1	0	6.0	1	6.0

TABLE 7.4. (*Continued*)

TRT	CD4	SEX	RACE	WT	HOMO	PCP	PDATE	DIE	DDATE
A	85	1	1	152	0	0	10.8	0	10.8
B	6	1	1	151	1	0	13.0	0	13.0
B	53	1	2	115	0	0	8.9	0	8.9
A	386	1	1	220	1	0	27.6	0	27.6
A	12	1	1	130	1	0	26.4	1	26.4
B	356	0	1	110	0	0	27.8	0	27.8
A	19	1	1	187	0	0	28.0	0	28.0
A	39	1	2	135	0	0	2.9	0	2.9
B	9	1	1	139	0	1	6.9	0	8.0
B	44	1	2	112	0	0	23.1	0	23.1
B	7	1	1	141	1	0	15.9	1	15.9
A	34	1	1	110	1	1	0.4	1	6.1
B	126	1	1	155	1	1	0.2	0	6.9
B	4	1	1	142	1	0	20.3	0	20.3
A	16	1	1	154	0	0	14.7	1	14.7
A	22	1	1	121	1	0	21.4	0	21.4
B	35	1	1	165	1	0	21.2	0	21.2
A	98	1	1	167	1	0	17.5	0	17.5
A	357	1	1	133	0	0	16.6	0	16.6
B	209	1	1	146	1	0	15.6	0	15.6
A	138	1	1	134	1	0	8.8	0	8.8
B	36	1	1	169	1	0	3.4	0	3.4
A	90	1	1	166	0	0	30.0	0	30.0
B	51	1	1	120	0	1	26.0	0	26.4
B	25	1	2	161	0	0	29.0	0	29.0
A	17	1	1	130	0	0	7.3	1	7.3
A	73	1	1	140	0	0	20.9	0	20.9
A	123	1	1	134	1	0	17.4	0	17.4
B	161	1	1	177	1	1	19.3	1	19.3
B	105	1	1	128	1	1	3.7	0	23.5
A	74	1	2	134	1	0	24.8	0	24.8
A	7	1	1	13	0	0	10.3	1	10.3
B	29	0	1	97	0	1	13.1	0	23.9
A	84	1	1	217	1	0	24.8	0	24.8
B	9	1	1	158	1	0	23.5	0	23.5
A	29	1	1	16	1	0	23.5	0	23.5
B	24	1	1	136	1	0	19.4	0	19.4
B	715	1	2	126	1	0	15.7	0	15.7
B	4	1	2	159	0	0	17.1	0	17.1
A	147	1	1	170	1	1	9.8	0	16.8
A	162	1	1	137	0	0	11.0	0	11.0
B	35	1	1	150	1	0	11.9	0	11.9
B	14	1	1	153	1	0	8.2	1	8.2

TABLE 7.4. (*Continued*)

TRT	CD4	SEX	RACE	WT	HOMO	PCP	PDATE	DIE	DDATE
B	227	1	1	150	1	0	9.5	0	9.5
B	137	1	1	145	1	0	9.0	0	9.0
A	48	1	1	143	0	0	8.3	0	8.3
A	62	1	1	175	1	0	6.7	0	6.7
A	47	1	1	164	1	0	5.5	0	5.5
B	7	1	1	205	0	0	6.9	0	6.9
B	9	1	1	121	0	1	19.4	0	23.9
B	243	1	2	152	0	1	12.0	1	0.0
A	133	1	1	136	0	0	23.1	0	23.1
A	56	1	1	159	1	0	23.2	0	23.2
A	11	1	1	157	0	0	8.7	1	8.7
A	94	1	2	116	0	0	15.1	1	15.1
A	68	1	1	185	1	0	21.3	0	21.3
A	139	1	1	145	1	0	19.1	0	19.1
B	15	0	1	114	0	0	17.4	0	17.4
B	22	1	2	125	1	0	4.4	1	4.4

- Patient characteristics: Baseline CD4 count, Sex (1 = Male), Race (1 = White, 2 = Black, and 3 = Other), Weight (lbs), and Homosexuality (1 = Yes, 0 = No)
- PCP recurrence indicator (1 = Yes) and PDATE or time to recurrence (months)
- DIE or Survival indicator (1 = Yes), and DDATE or time to death (or to date last seen for survivors; in months)

Consider each of these endpoints: relapse (treating death as censoring), death (treating relapse as censoring), and death or relapse (which ever comes first). For each endpoint, answer these questions:

(a) Estimate the survival function for white, homosexual men.

(b) Estimate the survival functions, one for each treatment.

(c) Compare the two treatments; is the treatment different short-term or long-term?

(d) Compare men versus women.

(e) Taken collectively, do the covariates contribute significantly to prediction of survival?

(f) Fit the multiple regression model to obtain estimates of individual regression coefficients and their standard errors. Draw your conclusion concerning the conditional contribution of each factor.

(g) Within the context of the multiple regression model in (b), does treatment alter the effect of CD4?

(h) Focusing on treatment as the primary factor, taken collectively, was this main effect altered by any other covariates?

(i) Within the context of the multiple regression model in (b), is the effect of CD4 linear?

(j) Do treatment and CD4, individually, fit the proportional hazards model?

Bibliography

Agresti, A. (1990). *Categorical Data Analysis*. New York: John Wiley and Sons.

Ahlquist, D. A., D. B. McGill, S. Schwartz, and W. F. Taylor, (1985). Fecal blood levels in health and disease: A study using HemoQuant. *New England Journal of Medicine* 314: 1422.

Bamber, D. (1975). The area above the ordinal dominance graph and the area below the receiver operating graph. *Journal of Mathematical Psychology* 12: 387–415.

Begg, C. B. and B. McNeil (1988). Assessment of radiologic tests: Control of bias and other design considerations. *Radiology* 167: 565–569.

Berkowitz, G. S. (1981). An epidemiologic study of preterm delivery. *American Journal of Epidemiology* 113: 81–92.

Berry, G. (1983). The analysis of mortality by the subject-years method. *Biometrics* 39: 173–184.

Bhattacharyya, G. K., M. G. Karandinos, and G. R. DeFoliart (1979). Point estimates and confidence intervals for infection rates using pooled organisms in epidemiologic studies. *American Journal of Epidemiology* 109: 124–131.

Bishop, Y. M. M., S. E. Fienberg, and P. W. Holland (1975). *Discrete Multivariate Analyses: Theory and Practice*. Cambridge, MA: MIT Press.

Blot, W. J., M. Harrington, A. Toledo, R. Hoover, C. W. Heath, and J. F. Fraumeni (1978). Lung cancer after employment in shipyards during World War II. *New England Journal of Medicine* 299: 620–624.

Breslow, N. (1970). A generalized Kruskal–Wallis test for comparing *K* samples subject to unequal patterns of censorship. *Biometrika* 57: 579–594.

Breslow, N. (1982). Covariance adjustment of relative-risk estimates in matched studies. *Biometrics* 38: 661–672.

Breslow, N. E. and N. E. Day (1980). *Statistical Methods in Cancer Research, Vol. I: The Analysis of Case-Control Studies*. Lyons: International Agency for Research on Cancer.

Brown, B. W. (1980). Prediction analyses for binary data. In: *Biostatistics Casebook*, edited by R. G. Miller, B. Efron, B. W. Brown, and L. E. Moses. New York: John Wiley & Sons, pp. 3–18.

Chin, T., W. Marine, E. Hall, C. Gravelle, and J. Speers (1961). The influence of Salk vaccination on the epidemic pattern and the spread of the virus in the community. *American Journal of Hygiene* 73: 67–94.

Cohen, J. (1960). A coefficient of agreement for nominal scale. *Educational and Psychological Measurements* 20: 37–46.

Connett, J. E., F. Rhame, J. Thomas, and C. T. Le (1990). Estimation of infectious potential of blood containing human immunodeficiency virus. *Biometrical Journal* 32: 781–789.

Conover, W. J. (1974). Some reasons for not using the Yate continuity correction in 2×2 contingency tables. *Journal of the American Statistical Association* 69: 374–378.

Cox, D. R. (1972). Regression models and life tables. *Journal of the Royal Statistical Society, Series B* 34: 187–220.

Cox, D. R. and D. Oakes (1984). *Analysis of Survival Data*. New York: Chapman and Hall.

Cox, D. R. and E. J. Snell (1989). *The Analysis of Binary Data*. London: Chapman and Hall, 2nd edition.

D'Angelo, L. J., J. C. Hierholzer, R. C. Holman, and J. D. Smith (1981). Epidemic keratoconjunctivitis caused by adenovirus Type 8: Epidemiologic and laboratory aspects of a large outbreak. *American Journal of Epidemiology* 113: 44–49.

Daniel, W. W. (1987). *Biostatistics: A Foundation for Analysis in the Health Sciences*. New York: John Wiley and Sons.

Efron, B. (1978). Regression and ANOVA with zero-one data: Measures of residual variation. *Journal of the American Association* 73: 113–121.

Freeman, D. H. (1980). *Applied Categorical Data Analysis*. New York: Marcel Dekker.

Freireich, E. J., E. Gehan, E. Frei, III, L. R. Schroeder, I. J. Wolman, R. Anbari, E. O. Burgert, S. D. Mills, D. Pinkel, O. S. Selawry, J. H. Moon, B. R. Gendel, C. L. Spurr, R. Storrs, F. Haurani, B. Hoogstraten, and S. Lee (1963). The effect of 6-mercaptoburine on the duration of steroid-induced remissions in acute leukemia: A model for evaluation of other potentially useful therapy. *Blood* 21: 699–716.

Frome, E. L. (1983). The analysis of rates using Poisson regression models. *Biometrics* 39: 665–674.

Frome, E. L. and H. Checkoway (1985). Use of Poisson regression models in estimating rates and ratios. *American Journal of Epidemiology* 121: 309–323.

Gart, J. J. (1969). An exact test for comparing matched proportions in crossover designs. *Biometrika* 56: 75–80.

Gehan, E. A. (1965a). A generalized Wilcoxon test for comparing arbitrarily singly censored samples. *Biometrika* 52: 203–223.

Gehan, E. A. (1965b). A generalized two-sample Wilcoxon test for doubly censored data. *Biometrika* 52: 650–653.

Graham, S., J. Marshall, B. Haughey, A. Mittleman, M. Swanson, M. Zielezny, T. Byers, G. Wilkinson, and D. West (1988). Dietary epidemiology of cancer of the colon in western New York. *American Journal of Epidemiology* 128: 490–503.

Greenwood, M. (1926). The natural duration of cancer. *Reports on public health and medical subjects, Her Majesty's Stationery Office* 33: 1–26.

Hanley, J. A. and B. J. McNeil (1982). The meaning and use of the area under a receiver operating characteristic (ROC) curve. *Radiology* 143: 29–36.

Hanley, J. A. and B. J. McNeil (1983). Method for comparing the area under the ROC curves derived from the same cases. *Radiology* 148: 839.

Helsing, K. J. and M. Szklo (1981). Mortality after bereavement. *American Journal of Epidemiology* 114: 41–52.

Herbst, A. L., H. Ulfelder, and D. C. Poskanzer (1971). Adenocarcinoma of the vagina. *New England Journal of Medicine* 284: 878–881.

Hlatky, M. A., D. B. Pryor, F. E. Harrell, R. M. Califf, D. B. Mark, and R. A. Rosati (1984). Factors affecting sensitivity and specificity of exercise electrocardiography: Multivariate analysis. *American Journal of Medicine* 77: 64–71.

Holford, T. R. (1982). Cavariance analysis for case-control studies with small blocks. *Biometrics* 38: 673–683.

Hollows, F. C. and P. A. Graham (1966). Intraocular pressure, glaucoma, and glaucoma suspects in a defined population. *British Journal of Ophthalmology* 50: 570–586.

Hosmer, D. W. Jr., and S. Lemeshow (1989). *Applied Logistic Regression.* New York: John Wiley and Sons.

Jackson, R., R. Scragg, and R. Beaglehole (1992). Does recent alcohol consumption reduce the risk of acute myocardial infarction and coronary death in regular drinkers? *American Journal of Epidemiology* 136: 819–824.

Kaplan, E. L. and P. Meier (1958). Nonparametric estimation from incomplete observations. *Journal of the American Statistical Association* 53: 457–481.

Kelsey, J. L., V. A. Livolsi, T. R. Holford, D. B. Fischer, E. D. Mostow, P. E. Schartz, T. O'Connor, and C. White (1982). A case-control study of cancer of the endometrium. *American Journal of Epidemiology* 116: 333–342.

Kleinbaum, D. G., L. L. Kupper, and K. E. Muller (1988). *Applied Regression Analysis and Other Multivariate Methods*. Boston: PWS-Kent Publishing.

Kleinman, J. C. and A. Kopstein (1981). Who is being screened for cervical cancer? *American Journal of Public Health* 71: 73–76.

Korff, F. A., M. A. M. Taback, and J. H. Beard (1952). A coordinated investigation of a food poissoning outbreak. *Public Health Reports* 67: 909–913.

Le, C. T. (1981). A new estimator for infection rates using pools of variable size. *American Journal of Epidemiology* 114: 132–135.

Le, C. T. (1997a). *Applied Survival Analysis*. New York: John Wiley and Sons.

Le, C. T. (1997b). Evaluation of confounding effects in ROC studies. *Biometrics* 53: 998–1007.

Le, C. T. and B. R. Lindgren (1988). Computational implementation of the conditional logistic regression model in the analysis of epidemiologic matched studies. *Computers and Biomedical Research* 21: 48–52.

Le, C. T., K. A. Daly, R. H. Margolis, B. R. Lindgren, and G. S. Giebink (1992). A clinical profile of otitis media. *Archives of Otolaryngology* 118: 1225–1228.

Li, D. K., J. R. Daling, A. S. Stergachis, J. Chu, and N. S. Weiss (1990). Prior condom use and the risk of tubal pregnancy. *American Journal of Public Health* 80: 964–966.

Liang, K. Y. and S. G. Self (1985). Tests for homogeneity of odds ratio when data are sparse. *Biometrika* 72: 353–358.

Mack, T. M., M. C. Pike, B. E. Henderson, R. I. Pfeffer, V. R. Gerkins, M. Arthur, and S. E. Brown (1976). Estrogens and endometrial cancer in a retirement community. *New England Journal of Medicine* 294: 1262–1267.

Mantel, N., and W. Haenszel (1959). Statistical aspects of the analysis of data from retrospective studies of disease. *Journal of the National Cancer Institute* 22: 719–748.

Martinez, F. D., A. L. Wright, C. J. Holber, W. J. Morgan, and L. M. Taussig (1992). Maternal age as a risk factor for wheezing lower respiratory illness in the first year of life. *American Journal of Epidemiology* 136: 1258–1268.

May, D. (1974). Error rates in cervical cytological screening tests. *British Journal of Cancer* 29: 106–113.

McCullagh, P. (1980). Regression models for ordinal data. *Journal of the Royal Statistical Society-B* 42: 109–142.

McFadden, D. (1974). Conditional logit analysis of qualitative choice behavior. In: *Frontiers in Econometrics*, edited by Zarembka. New York: Academic Press, pp. 105–142.

Murray, D. M., C. L. Perry, C. O'Connell, and L. Schmid (1987). Seventh-grade cigarette, alcohol, and marijuana use: Distribution in a north central U.S. metropolitan population. *International Journal of the Addictions* 22: 357–376.

Negri, E., C. L. Vecchia, P. Bruzzi, G. Dardonovi, A. Decali, D. Palli, F. Parazzini, and M. R. Del Turco (1988). Risk factors for breast cancer: Pooled results from three Italian case-control studies. *American Journal of Epidemiology* 128: 1207–1215.

Nischan, P., K. Ebeling, and C. Schindler (1988). Smoking and invasive cervical cancer risk: Results from a case-control study. *American Journal of Epidemiology* 128: 74–77.

Padian, N. S. (1990). Sexual histories of heterosexual couples with one HIV-infected partner. *American Journal of Public Health* 80: 990–991.

Peto, R. (1972). Contribution to discussion of paper by D. R. Cox. *Journal of the Royal Statistical Society, Series B*, 34: 205–207.

Rosenberg, L., D. Slone, S. Shapiro, D. W. Kaufman, and O. S. Miettinen (1981). Case-control studies on the acute effects of coffee upon the risk of myocardial infarction: Problems in the selection of a hospital control series. *American Journal of Epidemiology* 113: 646–652.

Rosner, B. (1982). Statistical methods in ophthalmology: An adjustment for the intra-class correlation between eyes. *Biometrics* 38: 105–114.

Rousch G. C., J. A. Kelly, J. W. Meigs, and J. T. Flannery (1982). Scrotal carcinoma in Connecticut metal workers: Sequel to a study of sinonasal cancer. *American Journal of Epidemiology* 116: 76–85.

Schwarts, B. S., R. L. Doty, C. Monroe, R. Frye, and S. Barker (1989). Olfactory function in chemical workers exposed to acrylate and methacrylate vapors. *American Journal of Public Health* 79: 613–618.

Shapiro, S., L. Rosenberg, D. Slone, and D. W. Kaufman (1979). Oral contraceptive use in relation to myocardial infarction. *Lancet* 1: 743–746.

Strader, C. H., W. S. Weiss, and J. R. Daling (1988). Vasectomy and the incidence of testicular cancer. *American Journal of Epidemiology* 128: 56–63.

Strogatz, D. (1990). Use of medical care for chest pain differences between blacks and whites. *American Journal of Public Health* 80: 290–293.

Stuart, A. (1955). A test for homogeneity of marginal distribution in two-way classification. *Biometrika* 42: 412–416.

Tarone, R. E. and J. Ware (1977). On distribution-free tests for equality of survival distributions. *Biometrika* 64: 156–160.

Tosteson, A. A. and C. B. Begg (1988). A general regression methodology for ROC-curve estimation. *Medical Decision Making* 8: 204–215.

True, W. R., J. Golberg, and S. A. Eisen (1988). Stress symptomology among Vietnam veterans. *American Journal of Epidemiology* 128: 85–92.

Tuyns A. J., G. Pequignot, and O. M. Jensen (1977). Esophageal cancer in Ille-et-Vilaine in relation to alcohol and tobacco consumption: Multiplicative risks. *Bulletin of Cancer* 64: 45–60.

Walter, S. D., S. W. Hildreth, and B. J. Beaty (1980). Estimation of infection rates in population of organisms using pools of variable size. *American Journal of Epidemiology* 112: 124–128.

Wermuth, N. (1976). Explanatory analyses of multidimensional contingency tables. *Proceedings of the 9th International Biometrics Conference* 1: 279–295.

Whittemore, A. S., R. Harris, J. Itnyre, and The Collaborative Ovarian Cancer Group (1992). Characteristics relating to ovarian cancer risk: Collaborative analysis of 12 U.S. case-control studies. *American Journal of Epidemiology* 136: 1184–1203.

Whittemore, A. S., M. L. Wu, R. S. Paffenbarger, Jr., D. L. Sarles, J. B. Kampert, S. Grosser, D. L. Jung, S. Ballon, and M. Hendrickson (1988). Personal and environmental characteristics related to epithelial ovarian cancer. *American Journal of Epidemiology* 128: 1228–1240.

Wise, M. E. (1954). A quickly convergent expansion for cumulative hypergeometric probabilities, direct and inverse. *Biometrika* 41: 317–329.

Yen, S., C. Hsieh, and B. MacMahon (1982). Consumption of alcohol and tobacco and other risk factors for pancreatitis. *American Journal of Epidemiology* 116: 407–414.

Index

WILEY SERIES IN PROBABILITY AND STATISTICS

ESTABLISHED BY WALTER A. SHEWHART AND SAMUEL S. WILKS

Probability and Statistics Section

*ANDERSON · The Statistical Analysis of Time Series
ARNOLD, BALAKRISHNAN, and NAGARAJA · A First Course in Order Statistics
ARNOLD, BALAKRISHNAN, and NAGARAJA · Records
BACCELLI, COHEN, OLSDER, and QUADRAT · Synchronization and Linearity: An Algebra for Discrete Event Systems
BASILEVSKY · Statistical Factor Analysis and Related Methods: Theory and Applications
BERNARDO and SMITH · Bayesian Statistical Concepts and Theory
BILLINGSLEY · Convergence of Probability Measures
BOROVKOV · Asymptotic Methods in Queuing Theory
BRANDT, FRANKEN, and LISEK · Stationary Stochastic Models
CAINES · Linear Stochastic Systems
CAIROLI and DALANG · Sequential Stochastic Optimization
CONSTANTINE · Combinatorial Theory and Statistical Design
COVER and THOMAS · Elements of Information Theory
CSÖRGÖ and HORVÁTH · Weighted Approximations in Probability Statistics
CSÖRGÖ and HORVÁTH · Limit Theorems in Change Point Analysis
DETTE and STUDDEN · The Theory of Canonical Moments with Applications in Statistics, Probability, and Analysis
*DOOB · Stochastic Processes
DRYDEN and MARDIA · Statistical Analysis of Shape
DUPUIS and ELLIS · A Weak Convergence Approach to the Theory of Large Deviations
ETHIER and KURTZ · Markov Processes: Characterization and Convergence
FELLER · An Introduction to Probability Theory and Its Applications, Volume 1, *Third Edition*, Revised; Volume II, *Second Edition*
FULLER · Introduction to Statistical Time Series, *Second Edition*
FULLER · Measurement Error Models
GELFAND and SMITH · Bayesian Computation
GHOSH, MUKHOPADHYAY, and SEN · Sequential Estimation
GIFI · Nonlinear Multivariate Analysis
GUTTORP · Statistical Inference for Branching Processes
HALL · Introduction to the Theory of Coverage Processes
HAMPEL · Robust Statistics: The Approach Based on Influence Functions
HANNAN and DEISTLER · The Statistical Theory of Linear Systems
HUBER · Robust Statistics
IMAN and CONOVER · A Modern Approach to Statistics
JUREK and MASON · Operator-Limit Distributions in Probability Theory
KASS and VOS · Geometrical Foundations of Asymptotic Inference

*Now available in a lower priced paperback edition in the Wiley Classics Library.

Probability and Statistics (Continued)

KAUFMAN and ROUSSEEUW · Finding Groups in Data: An Introduction to Cluster Analysis
KELLY · Probability, Statistics, and Optimization
LINDVALL · Lectures on the Coupling Method
McFADDEN · Management of Data in Clinical Trials
MANTON, WOODBURY, and TOLLEY · Statistical Applications Using Fuzzy Sets
MARDIA and JUPP · Statistics of Directional Data, *Second Edition*
MORGENTHALER and TUKEY · Configural Polysampling: A Route to Practical Robustness
MUIRHEAD · Aspects of Multivariate Statistical Theory
OLIVER and SMITH · Influence Diagrams, Belief Nets and Decision Analysis
*PARZEN · Modern Probability Theory and Its Applications
PRESS · Bayesian Statistics: Principles, Models, and Applications
PUKELSHEIM · Optimal Experimental Design
RAO · Asymptotic Theory of Statistical Inference
RAO · Linear Statistical Inference and Its Applications, *Second Edition*
*RAO and SHANBHAG · Choquet-Deny Type Functional Equations with Applications to Stochastic Models
ROBERTSON, WRIGHT, and DYKSTRA · Order Restricted Statistical Inference
ROGERS and WILLIAMS · Diffusions, Markov Processes, and Martingales, Volume I: Foundations, *Second Edition;* Volume II: Îto Calculus
RUBINSTEIN and SHAPIRO · Discrete Event Systems: Sensitivity Analysis and Stochastic Optimization by the Score Function Method
RUZSA and SZEKELY · Algebraic Probability Theory
SCHEFFE · The Analysis of Variance
SEBER · Linear Regression Analysis
SEBER · Multivariate Observations
SEBER and WILD · Nonlinear Regression
SERFLING · Approximation Theorems of Mathematical Statistics
SHORACK and WELLNER · Empirical Processes with Applications to Statistics
SMALL and McLEISH · Hilbert Space Methods in Probability and Statistical Inference
STAPLETON · Linear Statistical Models
STAUDTE and SHEATHER · Robust Estimation and Testing
STOYANOV · Counterexamples in Probability
TANAKA · Time Series Analysis: Nonstationary and Noninvertible Distribution Theory
THOMPSON and SEBER · Adaptive Sampling
WELSH · Aspects of Statistical Inference
WHITTAKER · Graphical Models in Applied Multivariate Statistics
YANG · The Construction Theory of Denumerable Markov Processes

Applied Probability and Statistics Section

ABRAHAM and LEDOLTER · Statistical Methods for Forecasting
AGRESTI · Analysis of Ordinal Categorical Data
AGRESTI · Categorical Data Analysis
ANDERSON, AUQUIER, HAUCK, OAKES, VANDAELE, and WEISBERG · Statistical Methods for Comparative Studies
ARMITAGE and DAVID (editors) · Advances in Biometry
*ARTHANARI and DODGE · Mathematical Programming in Statistics
ASMUSSEN · Applied Probability and Queues
*BAILEY · The Elements of Stochastic Processes with Applications to the Natural Sciences

*Now available in a lower priced paperback edition in the Wiley Classics Library.

Applied Probability and Statistics (Continued)

BARNETT and LEWIS · Outliers in Statistical Data, *Third Edition*
BARTHOLOMEW, FORBES, and McLEAN · Statistical Techniques for Manpower Planning, *Second Edition*
BATES and WATTS · Nonlinear Regression Analysis and Its Applications
BECHHOFER, SANTNER, and GOLDSMAN · Design and Analysis of Experiments for Statistical Selection, Screening, and Multiple Comparisons
BELSLEY · Conditioning Diagnostics: Collinearity and Weak Data in Regression
BELSLEY, KUH, and WELSCH · Regression Diagnostics: Identifying Influential Data and Sources of Collinearity
BHAT · Elements of Applied Stochastic Processes, *Second Edition*
BHATTACHARYA and WAYMIRE · Stochastic Processes with Applications
BIRKES and DODGE · Alternative Methods of Regression
BLOOMFIELD · Fourier Analysis of Time Series: An Introduction
BOLLEN · Structural Equations with Latent Variables
BOULEAU · Numerical Methods for Stochastic Processes
BOX · Bayesian Inference in Statistical Analysis
BOX and DRAPER · Empirical Model-Building and Response Surfaces
BOX and DRAPER · Evolutionary Operation: A Statistical Method for Process Improvement
BUCKLEW · Large Deviation Techniques in Decision, Simulation, and Estimation
BUNKE and BUNKE · Nonlinear Regression, Functional Relations and Robust Methods: Statistical Methods of Model Building
CHATTERJEE and HADI · Sensitivity Analysis in Linear Regression
CHOW and LIU · Design and Analysis of Clinical Trials: Concepts and Methodologies
CLARKE and DISNEY · Probability and Random Processes: A First Course with Applications, *Second Edition*
*COCHRAN and COX · Experimental Designs, *Second Edition*
CONOVER · Practical Nonparametric Statistics, *Second Edition*
CORNELL · Experiments with Mixtures, Designs, Models, and the Analysis of Mixture Data, *Second Edition*
*COX · Planning of Experiments
CRESSIE · Statistics for Spatial Data, *Revised Edition*
DANIEL · Applications of Statistics to Industrial Experimentation
DANIEL · Biostatistics: A Foundation for Analysis in the Health Sciences, *Sixth Edition*
DAVID · Order Statistics, *Second Edition*
*DEGROOT, FIENBERG, and KADANE · Statistics and the Law
DODGE · Alternative Methods of Regression
DOWDY and WEARDEN · Statistics for Research, *Second Edition*
DUNN and CLARK · Applied Statistics: Analysis of Variance and Regression, *Second Edition*
ELANDT-JOHNSON and JOHNSON · Survival Models and Data Analysis
EVANS, PEACOCK, and HASTINGS · Statistical Distributions, *Second Edition*
FLEISS · The Design and Analysis of Clinical Experiments
FLEISS · Statistical Methods for Rates and Proportions, *Second Edition*
FLEMING and HARRINGTON · Counting Processes and Survival Analysis
GALLANT · Nonlinear Statistical Models
GLASSERMAN and YAO · Monotone Structure in Discrete-Event Systems
GNANADESIKAN · Methods for Statistical Data Analysis of Multivariate Observations, *Second Edition*
GOLDSTEIN and LEWIS · Assessment: Problems, Development, and Statistical Issues
GREENWOOD and NIKULIN · A Guide to Chi-Squared Testing
*HAHN · Statistical Models in Engineering
HAHN and MEEKER · Statistical Intervals: A Guide for Practitioners

*Now available in a lower priced paperback edition in the Wiley Classics Library.

Applied Probability and Statistics (Continued)

HAND · Construction and Assessment of Classification Rules
HAND · Discrimination and Classification
HEIBERGER · Computation for the Analysis of Designed Experiments
HINKELMAN and KEMPTHORNE: · Design and Analysis of Experiments, Volume 1: Introduction to Experimental Design
HOAGLIN, MOSTELLER, and TUKEY · Exploratory Approach to Analysis of Variance
HOAGLIN, MOSTELLER, and TUKEY · Exploring Data Tables, Trends and Shapes
HOAGLIN, MOSTELLER, and TUKEY · Understanding Robust and Exploratory Data Analysis
HOCHBERG and TAMHANE · Multiple Comparison Procedures
HOCKING · Methods and Applications of Linear Models: Regression and the Analysis of Variables
HOGG and KLUGMAN · Loss Distributions
HOLLANDER and WOLFE · Nonparametric Statistical Methods
HOSMER and LEMESHOW · Applied Logistic Regression
HØYLAND and RAUSAND · System Reliability Theory: Models and Statistical Methods
HUBERTY · Applied Discriminant Analysis
JACKSON · A User's Guide to Principle Components
JOHN · Statistical Methods in Engineering and Quality Assurance
JOHNSON · Multivariate Statistical Simulation
JOHNSON and KOTZ · Distributions in Statistics
Continuous Multivariate Distributions
JOHNSON, KOTZ, and BALAKRISHNAN · Continuous Univariate Distributions, Volume 1, *Second Edition*
JOHNSON, KOTZ, and BALAKRISHNAN · Continuous Univariate Distributions, Volume 2, *Second Edition*
JOHNSON, KOTZ, and BALAKRISHNAN · Discrete Multivariate Distributions
JOHNSON, KOTZ, and KEMP · Univariate Discrete Distributions, *Second Edition*
JUREČKOVÁ and SEN · Robust Statistical Procedures: Aymptotics and Interrelations
KADANE · Bayesian Methods and Ethics in a Clinical Trial Design
KADANE AND SCHUM · A Probabilistic Analysis of the Sacco and Vanzetti Evidence
KALBFLEISCH and PRENTICE · The Statistical Analysis of Failure Time Data
KELLY · Reversability and Stochastic Networks
KHURI, MATHEW, and SINHA · Statistical Tests for Mixed Linear Models
KLUGMAN, PANJER, and WILLMOT · Loss Models: From Data to Decisions
KLUGMAN, PANJER, and WILLMOT · Solutions Manual to Accompany Loss Models: From Data to Decisions
KOVALENKO, KUZNETZOV, and PEGG · Mathematical Theory of Reliability of Time-Dependent Systems with Practical Applications
LAD · Operational Subjective Statistical Methods: A Mathematical, Philosophical, and Historical Introduction
LANGE, RYAN, BILLARD, BRILLINGER, CONQUEST, and GREENHOUSE · Case Studies in Biometry
LAWLESS · Statistical Models and Methods for Lifetime Data
LEE · Statistical Methods for Survival Data Analysis, *Second Edition*
LEPAGE and BILLARD · Exploring the Limits of Bootstrap
LINHART and ZUCCHINI · Model Selection
LITTLE and RUBIN · Statistical Analysis with Missing Data
MAGNUS and NEUDECKER · Matrix Differential Calculus with Applications in Statistics and Econometrics
MALLER and ZHOU · Survival Analysis with Long Term Survivors
MANN, SCHAFER, and SINGPURWALLA · Methods for Statistical Analysis of Reliability and Life Data

*Now available in a lower priced paperback edition in the Wiley Classics Library.

Applied Probability and Statistics (Continued)

McLACHLAN and KRISHNAN · The EM Algorithm and Extensions
McLACHLAN · Discriminant Analysis and Statistical Pattern Recognition
McNEIL · Epidemiological Research Methods
MEEKER and ESCOBAR · Statistical Methods for Reliability Data
MILLER · Survival Analysis
MONTGOMERY and PECK · Introduction to Linear Regression Analysis, *Second Edition*
MYERS and MONTGOMERY · Response Surface Methodology: Process and Product in Optimization Using Designed Experiments
NELSON · Accelerated Testing, Statistical Models, Test Plans, and Data Analyses
NELSON · Applied Life Data Analysis
OCHI · Applied Probability and Stochastic Processes in Engineering and Physical Sciences
OKABE, BOOTS, and SUGIHARA · Spatial Tesselations: Concepts and Applications of Voronoi Diagrams
PANKRATZ · Forecasting with Dynamic Regression Models
PANKRATZ · Forecasting with Univariate Box-Jenkins Models: Concepts and Cases
PIANTADOSI · Clinical Trials: A Methodologic Perspective
PORT · Theoretical Probability for Applications
PUTERMAN · Markov Decision Processes: Discrete Stochastic Dynamic Programming
RACHEV · Probability Metrics and the Stability of Stochastic Models
RÉNYI · A Diary on Information Theory
RIPLEY · Spatial Statistics
RIPLEY · Stochastic Simulation
ROUSSEEUW and LEROY · Robust Regression and Outlier Detection
RUBIN · Multiple Imputation for Nonresponse in Surveys
RUBINSTEIN · Simulation and the Monte Carlo Method
RUBINSTEIN and MELAMED · Modern Simulation and Modeling
RYAN · Statistical Methods for Quality Improvement
SCHUSS · Theory and Applications of Stochastic Differential Equations
SCOTT · Multivariate Density Estimation: Theory, Practice, and Visualization
*SEARLE · Linear Models
SEARLE · Linear Models for Unbalanced Data
SEARLE, CASELLA, and McCULLOCH · Variance Components
STOYAN, KENDALL, and MECKE · Stochastic Geometry and Its Applications, *Second Edition*
STOYAN and STOYAN · Fractals, Random Shapes and Point Fields: Methods of Geometrical Statistics
THOMPSON · Empirical Model Building
THOMPSON · Sampling
TIJMS · Stochastic Modeling and Analysis: A Computational Approach
TIJMS · Stochastic Models: An Algorithmic Approach
TITTERINGTON, SMITH, and MAKOV · Statistical Analysis of Finite Mixture Distributions
UPTON and FINGLETON · Spatial Data Analysis by Example, Volume I: Point Pattern and Quantitative Data
UPTON and FINGLETON · Spatial Data Analysis by Example, Volume II: Categorical and Directional Data
VAN RIJCKEVORSEL and DE LEEUW · Component and Correspondence Analysis
WEISBERG · Applied Linear Regression, *Second Edition*
WESTFALL and YOUNG · Resampling-Based Multiple Testing: Examples and Methods for p-Value Adjustment
WHITTLE · Systems in Stochastic Equilibrium
WOODING · Planning Pharmaceutical Clinical Trials: Basic Statistical Principles

*Now available in a lower priced paperback edition in the Wiley Classics Library.

WOOLSON · Statistical Methods for the Analysis of Biomedical Data
*ZELLNER · An Introduction to Bayesian Inference in Econometrics

Texts and References Section

AGRESTI · An Introduction to Categorical Data Analysis
ANDERSON · An Introduction to Multivariate Statistical Analysis, *Second Edition*
ANDERSON and LOYNES · The Teaching of Practical Statistics
ARMITAGE and COLTON · Encyclopedia of Biostatistics: Volumes 1 to 6 with Index
BARTOSZYNSKI and NIEWIADOMSKA-BUGAJ · Probability and Statistical Inference
BERRY, CHALONER, and GEWEKE · Bayesian Analysis in Statistics and Econometrics: Essays in Honor of Arnold Zellner
BHATTACHARYA and JOHNSON · Statistical Concepts and Methods
BILLINGSLEY · Probability and Measure, *Second Edition*
BOX · R. A. Fisher, the Life of a Scientist
BOX, HUNTER, and HUNTER · Statistics for Experimenters: An Introduction to Design, Data Analysis, and Model Building
BOX and LUCEÑO · Statistical Control by Monitoring and Feedback Adjustment
BROWN and HOLLANDER · Statistics: A Biomedical Introduction
CHATTERJEE and PRICE · Regression Analysis by Example, *Second Edition*
COOK and WEISBERG · An Introduction to Regression Graphics
COX · A Handbook of Introductory Statistical Methods
DILLON and GOLDSTEIN · Multivariate Analysis: Methods and Applications
DODGE and ROMIG · Sampling Inspection Tables, *Second Edition*
DRAPER and SMITH · Applied Regression Analysis, *Third Edition*
DUDEWICZ and MISHRA · Modern Mathematical Statistics
DUNN · Basic Statistics: A Primer for the Biomedical Sciences, *Second Edition*
FISHER and VAN BELLE · Biostatistics: A Methodology for the Health Sciences
FREEMAN and SMITH · Aspects of Uncertainty: A Tribute to D. V. Lindley
GROSS and HARRIS · Fundamentals of Queueing Theory, *Third Edition*
HALD · A History of Probability and Statistics and their Applications Before 1750
HALD · A History of Mathematical Statistics from 1750 to 1930
HELLER · MACSYMA for Statisticians
HOEL · Introduction to Mathematical Statistics, *Fifth Edition*
JOHNSON and BALAKRISHNAN · Advances in the Theory and Practice of Statistics: A Volume in Honor of Samuel Kotz
JOHNSON and KOTZ (editors) · Leading Personalities in Statistical Sciences: From the Seventeenth Century to the Present
JUDGE, GRIFFITHS, HILL, LÜTKEPOHL, and LEE · The Theory and Practice of Econometrics, *Second Edition*
KHURI · Advanced Calculus with Applications in Statistics
KOTZ and JOHNSON (editors) · Encyclopedia of Statistical Sciences: Volumes 1 to 9 wtih Index
KOTZ and JOHNSON (editors) · Encyclopedia of Statistical Sciences: Supplement Volume
KOTZ, REED, and BANKS (editors) · Encyclopedia of Statistical Sciences: Update Volume 1
KOTZ, REED, and BANKS (editors) · Encyclopedia of Statistical Sciences: Update Volume 2
LAMPERTI · Probability: A Survey of the Mathematical Theory, *Second Edition*
LARSON · Introduction to Probability Theory and Statistical Inference, *Third Edition*
LE · Applied Categorical Data Analysis
LE · Applied Survival Analysis
MALLOWS · Design, Data, and Analysis by Some Friends of Cuthbert Daniel

*Now available in a lower priced paperback edition in the Wiley Classics Library.

Texts and References (Continued)

MARDIA · The Art of Statistical Science: A Tribute to G. S. Watson
MASON, GUNST, and HESS · Statistical Design and Analysis of Experiments with Applications to Engineering and Science
MURRAY · X-STAT 2.0 Statistical Experimentation, Design Data Analysis, and Nonlinear Optimization
PURI, VILAPLANA, and WERTZ · New Perspectives in Theoretical and Applied Statistics
RENCHER · Methods of Multivariate Analysis
RENCHER · Multivariate Statistical Inference with Applications
ROSS · Introduction to Probability and Statistics for Engineers and Scientists
ROHATGI · An Introduction to Probability Theory and Mathematical Statistics
RYAN · Modern Regression Methods
SCHOTT · Matrix Analysis for Statistics
SEARLE · Matrix Algebra Useful for Statistics
STYAN · The Collected Papers of T. W. Anderson: 1943–1985
TIERNEY · LISP-STAT: An Object-Oriented Environment for Statistical Computing and Dynamic Graphics
WONNACOTT and WONNACOTT · Econometrics, *Second Edition*

WILEY SERIES IN PROBABILITY AND STATISTICS

ESTABLISHED BY WALTER A. SHEWHART AND SAMUEL S. WILKS

Editors
Robert M. Groves, Graham Kalton, J. N. K. Rao, Norbert Schwarz, Christopher Skinner

Survey Methodology Section

BIEMER, GROVES, LYBERG, MATHIOWETZ, and SUDMAN · Measurement Errors in Surveys
COCHRAN · Sampling Techniques, *Third Edition*
COX, BINDER, CHINNAPPA, CHRISTIANSON, COLLEDGE, and KOTT (editors) · Business Survey Methods
*DEMING · Sample Design in Business Research
DILLMAN · Mail and Telephone Surveys: The Total Design Method
GROVES and COUPER · Nonresponse in Household Interview Surveys
GROVES · Survey Errors and Survey Costs
GROVES, BIEMER, LYBERG, MASSEY, NICHOLLS, and WAKSBERG · Telephone Survey Methodology
*HANSEN, HURWITZ, and MADOW · Sample Survey Methods and Theory, Volume I: Methods and Applications
*HANSEN, HURWITZ, and MADOW · Sample Survey Methods and Theory, Volume II: Theory
KASPRZYK, DUNCAN, KALTON, and SINGH · Panel Surveys
KISH · Statistical Design for Research
*KISH · Survey Sampling
LESSLER and KALSBEEK · Nonsampling Error in Surveys
LEVY and LEMESHOW · Sampling of Populations: Methods and Applications
LYBERG, BIEMER, COLLINS, de LEEUW, DIPPO, SCHWARZ, TREWIN (editors) · Survey Measurement and Process Quality
SKINNER, HOLT, and SMITH · Analysis of Complex Surveys

*Now available in a lower priced paperback edition in the Wiley Classics Library.

Printed in the United States
38249LVS00004B/1-96

9 780471 240600